George G. Hall

Molecular
Solid State Physics

With 29 Figures

Springer-Verlag
Berlin Heidelberg NewYork
London Paris Tokyo
Hong Kong Barcelona Budapest

Professor Dr. George G. Hall
Shell Centre for Mathematical Education
University of Nottingham
Nottingham NG7 2RD
United Kingdom

ISBN-13: 978-3-540-53792-2   e-ISBN-13: 978-3-642-84461-4
DOI: 10.1007/978-3-642-84461-4

51/3020-543210 Printed on acid-free paper.

Preface

This book originated from a course which I developed for the Master's degree course in Molecular Engineering in Kyoto University. Most of the students had degrees in Chemistry and a limited experience of Physics and Mathematics. Since research in Molecular Engineering requires knowledge of some applications of solid state physics which are not treated in conventional physics texts it was necessary to devise a course which would build on their chemical background and enable them to read the contemporary literature of relevance to their research. I hope that this book will be found useful as a text for other advanced courses on material science for chemists.

Molecular Engineering is concerned with the design and construction, at the molecular level, of materials which can fulfil specific functions. Thus the study of the forces between molecules and the influence of molecular shapes and electrostatic features on molecular properties are important. The mechanisms whereby, in the solid state, these produce cooperative effects, catalytic effects and abnormal electrical effects must be understood, at least qualitatively.

The aim of this book has been to give insight into the mechanisms whereby molecules influence one another when they are close together. This is of direct importance for many investigations within the present scope of the subject. The structures of solid materials under various conditions of temperature and pressure is one example. The modifications to chemical reactions when they take place on surfaces or in the body of a crystal and the flow of electrical charge through the material, are other examples. For this purpose certain theoretical ideas and techniques are essential but the demands on the reader's mathematical and physical background have been kept to a minimum. The text has been supplemented by Appendices, which relax this condition to some extent, so that a few important topics can be followed up in more detail.

A knowledge of these topics is also a first step in understanding the behaviour of biologically active molecules in the cell which is another concern of Molecular Engineering. Nevertheless all the materials considered here are dead. One of the missing factors can be tentatively identified as the system of the cell. The cell is a feed-back system that can use energy to access information and then use this information to control all its activities. In contrast, although many of the examples here involve the concept of information (in the form of entropy) it is not being controlled or changed from within the system. The objective of enlarging the subject to include this distinctive idea of quantum biology has, therefore, still to be achieved.

Another objective of Molecular Engineering has been the design of electronic materials on the molecular scale. The first stage in this must be an understanding of existing electronic components in terms of their electrons. The molecular computer, if it could be realized, would allow the miniaturization of computers to reach its natural limit. It would also allow the study of organisation and system to be realized in terms of known molecules in a new way.

The prospects of molecular solid state physics for the design of new materials as well as for molecular biology and for computing technology seem immense. I hope that the start made in this book will contribute to the realization of these long range objectives.

Acknowledgments

This book would never have been conceived without the stimulus of my Japanese colleagues. I am greatly in their debt for offering me the chair of Quantum Molecular Science and Technology in the new Division of Molecular Engineering at Kyoto University. In particular I must thank Professors Fukui, Nishijima, Yamabe, and Yonezawa who played major roles in my appointment. I am endebted to Professor Fujimoto who assisted me and guided me through many of the pitfalls of the Japanese lecture system. My thanks are also due to the many students who attended these lectures and commented on them.

My early University education has shaped in great measure my approach to this subject. I count myself fortunate to have had the experience of working with two of its pioneers. I owe much to the late Professors Ewald and Lennard-Jones. Their approach to the subject, with its emphasis on the clarity (and rigour) of the basic mathematical ideas and on the interplay between theory and experiment, has been my inspiration.

Finally, but fundamentally, I acknowledge the constant support and inspiration of my wife who encouraged this project from its inception and tolerated the disruption caused during its progress.

G G Hall

ホ
ー
ル

Contents

Contents

Chapter 7   Cooperative effects

Chapter 1   Close-Packed Crystals

1-1 Introduction

There are many factors which combine to determine the structure of a solid. In this chapter we shall consider one of the factors simplest to understand, namely the size and shape of the molecules composing a molecular solid. Crystals often achieve a packing of their molecules into the smallest possible volume. This geometrical factor determines completely the structure of some solids and packing considerations are always an important factor in fixing the shapes of all solids.

For our purpose, we require some simple theoretical techniques which will be of general use later. In particular, the use of scaling will be introduced. This technique can be applied only in certain special circumstances, which will be defined and which greatly limit its use, but it has the compensating advantage that its results are quite easy to obtain and yield much insight into the structure and properties of the solids. We also require a description of some of the very common crystal structures. There are many possible structures and a first classification in terms of the shape of the unit cell has been given so that other structures can be added in later chapters.

Since the crystals of the inert gases are composed of atoms, whose shapes are spherical, these are the simplest solids to use to illustrate the themes of this chapter. These are not the most exciting of crystals from a chemist's stand-point but they have been used so often as test-beds for new theoretical ideas that they have become interesting for that reason!

1-2 Some simple lattices

The easiest space lattice to visualise is the simple cubic lattice with its appearance like a tightly-packed collection of cubes. This structure is too simple to be the structure of any of these inert gas solids. The next simplest lattices are derived from this one. The face-centred cubic lattice (fcc) has extra atoms at the centre of each face of a cube as well as atoms at its corners. See Figure 1.

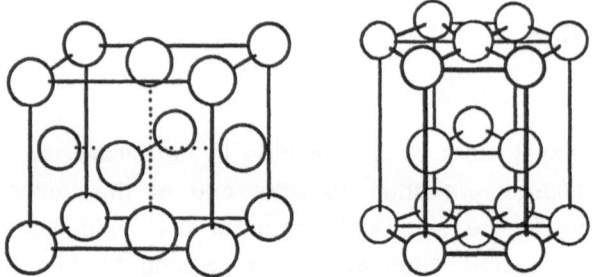

Fig.1 The fcc and hcp lattices

The body-centred cubic lattice (bcc) has atoms at the corners of a cubic lattice and extra atoms at the centre point of each cube. See Figure 2.

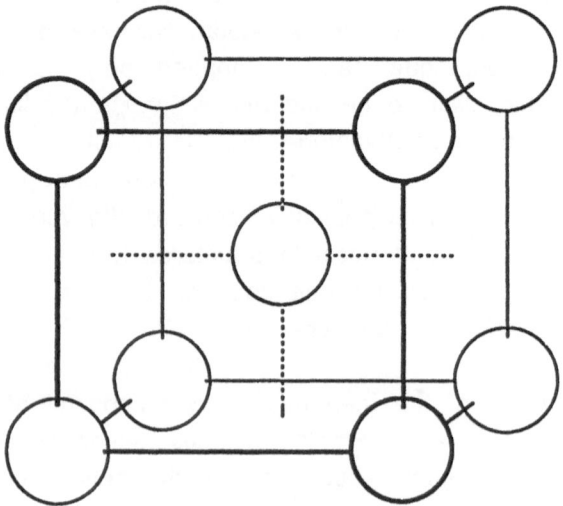

Fig. 2 The bcc lattice

It can be considered as two interlacing cubic lattices whose origins are separated by half the cube diagonal. The other lattice which is important for the discussion is the hexagonal close-packed lattice (hcp). See Figure 1. Its form is described more fully below.

In order to give practice in visualizing these lattices and because the property is important later we will consider one feature of these structures. A lattice will be called **alternant** if the lattice points can be divided into two classes so that no two members of the same class are nearest neighbours. (This term was introduced in relation to conjugated hydrocarbons but it is a topological term of general significance.) It is easy to verify that the simple cubic lattice is alternant by giving coordinates to each lattice point using one point as the origin and the sides of the cube as unit basis vectors. The coordinates of the lattice points then are all integers. The two classes comprise those points whose three coordinates add up to an even number and those whose sum is odd. Nearest neighbours of any given point always lie along one of the cube edges from that point and this implies that exactly one coordinate is increased or decreased by unity. Each class is itself a fcc lattice. Since the bcc lattice is composed of two similar cubic lattices, with the nearest neighbours of one lying on the other, its division into two classes is obvious and so it is alternant. On the other hand, some of the nearest neighbours of an atom in the fcc or hcp lattices are themselves nearest neighbours to one another so these lattices are not alternant.

To help us understand the reasons why particular solids adopt these structures we first consider the number of neighbours, $n(r)$, of a given lattice site at various distances of separation, r. These are shown in Table 1. It is convenient to use as the unit of distance, in this table, the distance between the nearest neighbours. The distances to higher order neighbours are listed as their squares. This table shows that the two lattices, the fcc and the hcp, have the same number, 12, of nearest neighbours and of next nearest neighbours, 6, but that they differ thereafter. The bcc lattice has fewer nearest neighbours but includes a greater total number of neighbours within a radius of twice the nearest distance.

Table 1. Number of neighbours at various distances in some lattices

fcc		bcc		hcp	
r^2	n	r^2	n	r^2	n
1	12	1	8	1	12
2	6	4/3	6	2	6
3	24	8/3	12	8/3	2
4	12	11/3	24	3	18
5	24	4	8	11/3	12
6	8	16/3	6	4	6

This table can be considered as the first step in creating a radial distribution function, g(r), for these crystals. This function is defined using a count of the number, N(r), of atomic nuclei within a sphere of radius r around a fixed nucleus. The distribution function is then defined formally as the derivative

$$g(r) = \frac{dN}{4\pi\rho r^2 dr}$$

where ρ is the average number density. Since the lattice structure implies a uniform density of atoms the asymptotic behaviour of g must approximate to the line g=1. For the ideal lattice and small values of r, g becomes a set of delta functions at the distances listed above and with weights given by n in Table 1. This is pictured, more realistically since the distances in the real crystal will be modified by the vibration of the atoms, as a set of bars with height proportional to n/r^2 in Figure 3.

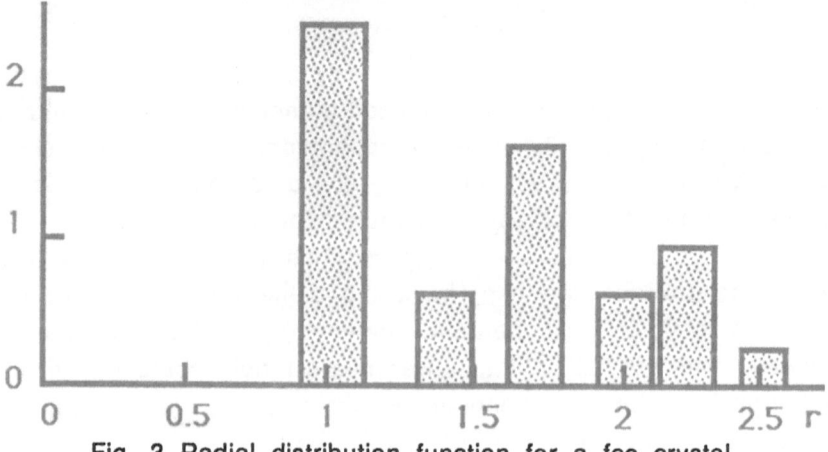

Fig. 3 Radial distribution function for a fcc crystal

1-3 The seven crystal systems

Although there is no intention of describing many of the 230 possible crystal structures it is useful to describe the seven systems into which they have been classified. (A complete description of all the structures is given by Zachariasen [1] and by Jaswon [2].) These are defined using the shape and size of the basic cell which is translated to produce the lattice. The three basis vectors are denoted by b_i, i=1,2,3, with lengths b_i, and the angles between them are α_i which is the angle between b_j and b_k, where all i,j,k are different.

The possible alternatives for the unit cell are shown in Table 2. The variables which can be determined freely are given first and then those which are fixed.

Table 2. The seven crystal systems

System	Basis set variables- free:	fixed
Triclinic	$b_1,b_2,b_3,\alpha_1,\alpha_2,\alpha_3$:	
Monoclinic	b_1,b_2,b_3,α_2:	$\alpha_1=\alpha_3=\pi/2$
Orthorhombic	b_1,b_2,b_3:	$\alpha_1=\alpha_2=\alpha_3=\pi/2$
Tetragonal	b_1,b_3:	$b_2=b_1,\alpha_1=\alpha_2=\alpha_3=\pi/2$
Trigonal	b_1,α_1:	$b_2=b_3=b_1,\ \alpha_1=\alpha_2=\alpha_3$
Hexagonal	b_1,b_3:	$b_2=b_1,\alpha_1=\alpha_2=\pi/2,\alpha_3=2\pi/3$
Cubic	b_1:	$b_2=b_3=b_1,\ \alpha_1=\alpha_2=\alpha_3=\pi/2$

The trigonal system contains two cases. The one listed above is the rhombohedral one. The second case is specified as under the hexagonal case. The difference is that it contains a three-fold symmetry axis rather than a six-fold symmetry axis. With these systems we have a first subdivision of the space groups into readily visualized categories.

1-4 The packing of spheres

It is a common experience, from the packing of fruit or marbles or other small spheres, that the closest packing of spheres in the plane is hexagonal with each sphere touching six nearest neighbours. On top of this layer of spheres can be placed a second hexagonal layer using the hollows of the first layer to reduce the vertical spacing. Only half of these hollows are occupied by this layer. When we come to the third layer there is a choice whether to place the spheres vertically above the positions of the spheres in the first layer, which, when repeated, produces the hexagonal close packed lattice, or to fill the hollows in the second layer which do not match the spheres in the first layer, and this, when repeated, produces the cubic close packed crystal (ccp, already defined as face centred cubic). Both structures give each atom twelve nearest neighbours and the volume of space occupied per atom is the same so they are equally examples of close packing. Both structures show planar close packing in various planes intersecting the layers so that the choice of layer is immaterial. Clearly these two structures are very similar physically and both are found in practice.

The simplest model of an inert gas atom is as a rigid sphere with a fixed radius. The radii vary from atom to atom. In addition there should be an energy reduction (i.e. greater stability) when two spheres touch. In the crystal these spheres will then pack together to have as much touching as possible. This leads to the two close-packed structures above (ccp and hcp) with 12 spheres touching each one. Actually, the He crystal has the hcp structure and the other inert gases have fcc crystals.

1-5 Forces between atoms

Although this rigid sphere model gives meaning to the size and shape of an atom it is not sufficiently detailed to allow many properties of a crystal to be evaluated. We now look at a better model which is more powerful. This starts from an analysis of the forces which act between atoms.

When two neutral atoms approach one another there is an instantaneous polarization of the electrons around one by any instantaneous distortion of the electrons around the other which produces a dipole moment. These fluctuations in dipole moment couple to give an attractive potential between the atoms which varies as an inverse sixth power of r, their distance apart. This dispersion, or van der Waals, force has long been familiar in discussions of intermolecular forces. When the atoms come closer a repulsion sets in to prevent them overlapping. This is less well understood and is represented by several alternative approximate formulae. For our purpose the most convenient form of repulsive potential is an inverse twelfth power. The resulting Lennard-Jones potential [3] has the form

$$u(r) = \varepsilon\left(\left(\frac{a}{r}\right)^{12} - 2\left(\frac{a}{r}\right)^{6}\right)$$

where ε is the depth of the attractive well and a is the value of r for which u is a minimum. This a is the distance of nearest approach of the two atoms and corresponds, roughly, to the sum of the radii of the two touching spheres. The significance of ε is that it is the energy bonus for two touching atoms. It is divided equally between the two atoms. The cohesive energy of the close-packed crystal (measured at 0°K), with 12 touching atoms, is then, approximately, 6ε per atom.

One advantage of this model is that the energy contribution of more distant neighbours can be estimated. The summation of these terms over the ccp lattice has been evaluated by Lennard-Jones and Ingham [4]. The resulting energy of an atom in association with all its neighbours is then:

$$U(r) = \frac{\varepsilon}{2}\left(12.13188 \left(\frac{a}{r}\right)^{12} - 28.90784 \left(\frac{a}{r}\right)^{6}\right)$$

The minimum value of this energy, with respect to the interatomic distance, is no longer at r=a but at the contracted value
$r_{min} = 0.97124$ a

and the cohesive energy per atom is increased to
$U(r_{min}) = 7.85836$ ε.

This demonstrates that the more distant neighbours do exert an appreciable influence on the quantitative results. The corresponding results for the hcp lattice are

r_{min} = 0.97123 a,

$U(r_{min})$ = 7.85893 ε.

The hcp, having a larger cohesive energy, should then be the more stable lattice for these crystals but by a very slight amount. In fact only the He crystal is hcp while the rest are ccp. The explanation for this has been sought by many authors and various alternative suggestions have been explored. The discussions are very tentative and inconclusive since the extra contributions, due to three-centre dispersion forces, exchange forces, charge transfer forces and to zero-point vibrational energy, are all so small.

1-6 Scaling of the potential

While it has been possible to give these results for the cohesive energy in explicit form it is not always possible to do this for other properties without considerable effort. There is, however, another method (following Lennard-Jones and Devonshire [5]) of obtaining results by scaling without much calculation. This starts by examining the form of the interatomic potential which is being used. This has the factor ε appearing in the formula and it is multiplied by a dimensionless term which has the value 1 when the two atoms are at their optimum separation. This suggests that we should arrange to measure all energies as multiples of ε and all distances as multiples of a. Volumes will then be multiples of $(a)^3$. The scaled potential then has exactly the same form, in terms of the scaled distance, for all the inert gas atoms and any property derived from it will have the same value in the appropriate scaled units. If we are interested in the relative values of properties for these solids then we can use this result extensively. We must be careful, however, to ensure that the property has no dependence on any other variable such as the mass of the atom or the external pressure or the external temperature. Several such properties can be found. As we have seen above, one of these is the cohesive energy, extrapolated to 0 K.

Another is the temperature of the melting point T_m since this transition point is also determined by the scaled potential. It will be easier, for our argument, to use RT_m since this is clearly an energy and so should be scaled with ε. The latent energy of melting, Λ, is also independent of external variables to a good approximation and its scaled value should be constant. As a test of this scaling procedure, Table 3 shows these properties (U and Λ in kcal/mole, T_m in K) except that RT_m has been used, instead of ε, to scale them so that only experimental properties are being used.

Table 3. Cohesive energies and latent heats of melting for inert gas solids

	T_m	U	U/RT_m	Λ	Λ/RT_m
Ne	24.6	0.45	9.2	0.080	1.638
Ar	83.2	1.85	11.2	0.2845	1.722
Kr	115.8	2.68	11.7	0.3917	1.703
Xe	161.4	3.82	11.9	0.5493	1.715

It is significant that the latent heat, scaled in this way, is the entropy of melting measured in dimensionless units. (This definition of entropy will be used throughout this book since it emphasises the connection with the concept of information.) Table 3 shows that the scaled properties are not exactly constant but only approximately so. This is partly because the potential function above is not an exact expression for the real potential and partly because there are other small contributions to the properties which do not scale in the same way. Scaling does not give absolute results but it does give, very simply, useful relative results. In fact, scaling may be a property of other forms of interatomic potential so that even perfect results from scaling could not establish this particular LJ form of potential. Conversely, the lack of strict constancy here shows the failure of any scaled potential to describe the situation exactly.

There are other properties which can be predicted, to a reasonable approximation, by scaling. One of these is α, the coefficient of thermal expansion of the crystal at a temperature just below its melting point. Since this is the relative expansion of the volume of the solid per degree rise in temperature it has the dimension of an inverse temperature and so the dimensionless product αT_m should be constant. These, using the compilation of Touloukian [6], are shown in Table 4. It is clear that this property is not as constant, when reduced, as are those above. This is partly due to the considerable variation in these properties in the neighbourhood of the melting point as well as to the simplicity of the model potential.

Table 4. Coefficient of volume expansion of a solid at its melting point

	α	αT_m
Ne	1.788	44.0
Ar	0.630	52.4
Kr	0.425	49.2
Xe	0.27	43.6

1-7 Surface energies

When a crystal is divided into two parts by pulling it apart at a plane there is a considerable loss of energy since the surface atoms lose half their neighbours. This loss can be evaluated using the interatomic potential energy expression above. There is then a second effect. The surface atoms are no longer in a minimum energy position. They will relax inwards under the attractive forces of those in the second layer down. This means that they will regain a small part of the lost energy. Again, the atoms in the second layer will be disturbed by this movement and will also move downwards but by a smaller amount. The disturbance may penetrate the crystal to several lower layers. This relaxation effect can also be calculated from the LJ potential.

These calculations have been performed by Benson and Claxton [7]. The surface energy is expressed as $\sigma = \sigma_0 + \Delta\sigma$, where the first term is the energy when the atoms are in fixed positions and the second term is the relaxation energy when all the positions are optimized. The contributions differ depending on which face of the crystal is exposed. Their results, in scaled or dimensionless form, are shown in Table 5. Each term has to be divided by ω, the area of unit cell on the surface, to give the specific energy and multiplied by ε to rescale it to the particular crystal.

Table 5. Surface energies for the ccp lattice

face	$-\omega\sigma_0/\varepsilon$	$\omega\Delta\sigma/\varepsilon$
{111}	0.37370	0.001233
{100}	0.44799	0.003958
{110}	0.66332	0.004475

These results show that the relaxation effect is very small. It is almost all due to the movement of the outer layer though even the tenth layer moves slightly. The {111} surfaces are the easiest to create.

1-8 Packing of ellipsoids

To enable us to consider the structure of the solids formed by the homonuclear diatomics we will extend our discussion from rigid spheres to rigid prolate ellipsoids. We start by packing these together in layers. Such ellipsoids can be packed together in a single layer with their long axes normal to the layer plane. They will then have contact at a circular cross-section so there will be six nearest neighbours touching every one, just as for spheres. On this layer another layer can be formed with its ends fitting into the gaps between molecules of the first layer and the structure will continue upwards as before. The result will be that in the vertical direction the unit cell of the lattice will be elongated because of the length of the major axis. The subsequent packing of layers may follow the hcp pattern of a two-layer repeat when the result will again be a hexagonal crystal or the ccp pattern of a three-layer repeat when it will be a monoclinic crystal.

It is possible, however, that the second layer may disturb the first in a more incisive way. If the molecules in the first layer remain parallel but all tilt a little to the normal, in the direction of a nearest neighbour, they can maintain contact but their centres may actually come closer to those in the second layer which are also tilted but in an opposite direction. This is an orthorhombic lattice. It is also possible that the tilt may take the molecular axis into the direction of a three-fold axis of the lattice. This produces a cubic lattice. These, and other more complicated lattices, are found experimentally for diatomics under various conditions of temperature and pressure. The most stable structures (at low temperatures) of H_2 and N_2 are cubic, O_2 is monoclinic while Cl_2, Br_2 and I_2 are orthorhombic. Since O_2 is anti-ferromagnetic it must also have magnetic forces acting which will modify its structure. The relative advantage of touching vertically or horizontally needs to be quantified to decide the relative merits of tilted and normal structures and this rigid model is not good enough to answer the question.

A full discussion of these solids has been given by English and Venables [8]. They show that a potential consisting of LJ potentials between each atom in one molecule and each atom in the other is required to model the size and shape of the diatomics. Thus the shape is more dumbbell-like than ellipsoidal. The ratio of the internuclear distance, d, to the value of σ, which they call the elongation, is the important dimensionless variable. The calculation of the cohesive energy of the various alternative structures then shows that the hexagonal form is most stable. Since this is not what is observed it becomes clear that another force is essential. The one which they introduce is the interaction between the quadrupole moments of the molecular charge densities. Since the quadrupole can vary from that of N_2 at $-1.5 \ 10^{-26}$ esu cm^2 through to that of F_2 at $+0.88 \ 10^{-26}$ esu cm^2, it is not surprising that this term has considerable effect in determining the relative energies of different structures. The angular dependence of this force means that the tilted structures are made more attractive relative to the normal ones. In particular, the cubic structure is strongly favoured by this interaction and can be made more stable by increasing the magnitude of the quadrupole moment. It then becomes the most stable structure for many values of the parameters. This helps to explain the structure of H_2 and N_2 which have relatively large quadrupole moments. For larger values of the elongation and smaller values of the quadrupole moment, as are found for the halogens, the orthorhombic structure becomes the most stable. The main features of the experimental situation have been reproduced by their theory although it is clear that the calculations are rather closely balanced so that small energy differences are not reliable. Since the LJ potential alone does not reproduce the correct result for the stability of the solid inert gases it cannot be expected to discriminate more accurately in this more complicated situation!

1-9 Sphere-like molecules and the plastic phase

The idea of spherical close packing extends naturally to the packing of molecules which have sufficiently high symmetry around their centres. Many of these molecules have an additional phase - the plastic phase - between the solid and the liquid. In this phase the centres of the molecules are fixed to lattice positions but the molecules exhibit rotational disorder. From their intermolecular distances it is clear that there is not enough space for free rotation of the molecules though some strongly coupled rotation may be possible. The disorder of the molecules makes them even more sphere-like. Thus the CH_4 molecular crystal has tetrahedral symmetry at low temperatures but it has a plastic phase which is a hcp structure. The adamantane molecule has icosahedral shape and it forms a ccp structure in its plastic phase.

These molecules, because of their centres of symmetry, have no dipole moment and very small electrostatic interactions and so their crystal structures are determined by the dispersion and short-range repulsion forces. Their crystals are generally low in cohesive energy and in entropy of melting.

The low-temperature solid phases of these molecules have various symmetries depending on the details of the packing of the molecules together. On the other hand the close packed structures predominate in the plastic phase so, at the transition point, there is often a change of structure. The extent of this phase is shown by the difference between the transition and the melting temperatures in Table 6. These molecules all have a ccp plastic phase.

Table 6. Transition and melting temperatures and entropies

	T_t	T_m	Λ_t/RT_t	Λ_m/RT_m
CCl_4	225.5	250.3	2.45	1.21
$C(CH_3)_3Cl$	219.0	248.6	4.38	1.01
c-Hexane	186.0	279.8	4.33	1.18
Camphor	250.0	453.0	3.83	1.41

The entropy of melting from the plastic phase to the liquid phase, which corresponds to that of the inert gas crystals in Table 1, has a value which is comparable. In the liquid phase the molecules become free to wander throughout the volume even though the number of nearest neighbours remains very similar. This freedom gives rise to a communal entropy change which can be estimated and accounts for 1 in these units. There will also be a contribution due to the increase in volume in the liquid. Thus the magnitude of these entropies is as expected. The entropy of the transition from the solid to the plastic phase is larger and more variable since there will be substantial changes in crystal geometry as well as the entropy due to the rotational disorder, which can be estimated as 1.5.

1-10 Origin of the repulsive term

The repulsion which arises when the charge clouds of two atoms begin to overlap can be traced to the operation of the Pauli Principle. This requires that the wavefunctions of the two atoms should combine into one antisymmetric wavefunction at small distances. With approximate wavefunctions this generally involves the introduction into the energy formulae of the overlap integrals between the basis functions on the two centres. A symmetrical method of doing this has been given by Löwdin [8]. The effect of this is to assign to the atom a modified shape which is no longer spherical. In the crystal this is important in the calculation of the elastic constants. It is as if the atom has been squashed by its neighbours. Since the atomic basis functions in use are exponential functions of distance the overlap integrals have an exponential dependence on the separation of the atoms. This is the origin of the exponential form of the repulsive potential. The inverse twelfth power is merely an approximation which is easier to handle.

Another approach to the problem has been given by Kim *et al* [10]. They have explored the possibility of using S, the overlap of the electron densities. They show that these overlaps are related to the accurately calculated repulsive potentials by the equation:

$$V_{rep} = A \, S^{\alpha}$$

where the exponent α is a little less than 1 and, with A, varies with the atoms concerned. Since these overlaps are easier to use than the overlaps of the orbitals they have become an alternative method of estimating repulsions for complicated molecules. The application of this idea to the interactions of diatomics has been given by Wheatley and Price [11]. A particular advantage of this method is that it relates easily to the electron density functional method of treating molecules. The conceptual picture of two molecules repelling because their electron distributions overlap is easy to accept.

References

[1] Zachariasen W H 1945 Theory of X-Ray Diffraction in Crystals, Wiley New York

[2] Jaswon M A 1965 Mathematical Crystallography, Longmans London

[3] Lennard-Jones J E 1925 Proc Roy Soc A106: 463

[4] Lennard-Jones J E and Ingham A E 1925 Proc Roy Soc A107: 636

[5] Lennard-Jones J E and Devonshire A F 1939 Proc Roy Soc A170: 464

[6] Touloukian Y S *et al* 1977 Thermal Expansion, IFI/Plenum New York

[7] Benson G C and Claxton T A 1964 J Phys Chem Solids 25: 367

[8] English C A and Venables J A 1974 Proc Roy Soc A340: 81

[9] Löwdin P O 1947 Arkiv Mat Astr Fysik 35A No 9, 1950 J Chem Phys 18: 365

[10] Kim Y S, Kim S K and Lee W D 1981 Chem Phys Lett 80 574

[11] Wheatley R J and Price S L 1990 Mol Phys 69: 507

Chapter 2   Ionic Crystals

2-1 Introduction

The crystal of rock salt was one of the first to have its structure determined using X-rays. It has a relatively simple geometrical form as have all the alkali halides. Salts which involve three or more different ions have more complicated structures. In this chapter we consider the forces which determine these structures and explore some crystal properties.

Because of the ionic charges, the ionic crystals involve strong electrostatic forces. These are long-range forces. They raise considerable difficulties of computation as well as of comprehension. Some of these problems are discussed and some are alluded to briefly. Long-range forces can give rise to new phenomena which depend on cooperative effects. Among these are the ferro-electricity of certain crystals and the high-temperature superconductivity of others. Cooperative effects are discussed more fully in Chapter 7.

Inevitably, in an introductory account, the quantum mechanical treatment of these crystals and their properties cannot be fully described. The interested reader is referred to Born and Huang [1] for a full account of the general theory and to Löwdin [2] for its detailed application to the structure and elastic properties of ionic crystals.

2-2 Alkali halide crystal structures

The NaCl structure is the one found most commonly among the alkali halides. It has the anions on a face-centred lattice and the cations on another face-centred lattice interlacing the first one. The two sets of ionic centres together would form a simple cubic lattice. The structure, therefore, makes use of the alternant property of the simple cubic lattice to ensure that each ion is surrounded by ions with the opposite charge. This structure is shown in Figure 1.

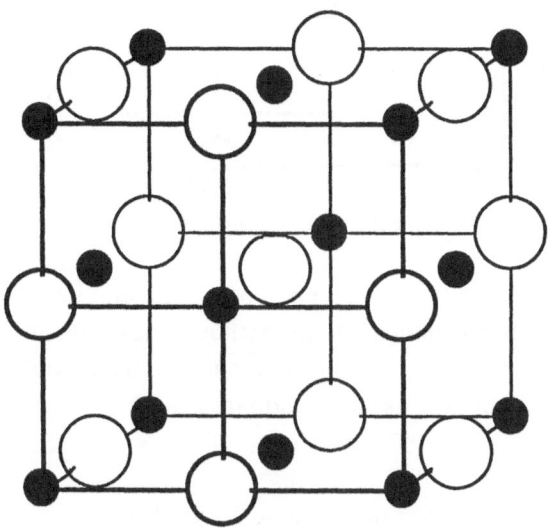

Fig.1 NaCl lattice

The other structure which is found among these crystals is the CsCl lattice structure. It has the anions on a simple cubic lattice and the cations are placed at the centre of each cube and form another cubic lattice. Together the ionic centres form a body-centred lattice and its alternant property is being used. This lattice is shown in Figure 2.

To describe the positions of the ions in these solids it is convenient to take one ion as the origin and use the cube edges to fix the direction of the axes and the distance between ions of the same sign along the axes as the unit of distance. This means that the ions at the centres of the faces, or the centre of the cube, will have fractional coordinates. It is not important for our purposes that these are not the correct basic lattice cells. Those cells are discussed in texts on crystallography.

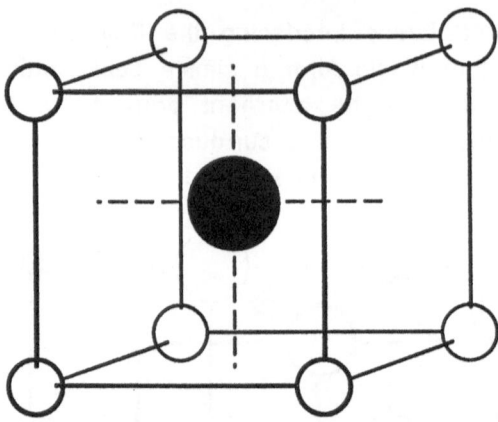

Fig. 2 CsCl lattice

The number and separation of neighbours in the two lattice structures are important in the calculation of properties. They are shown in Table 1. In this Table the unit of distance is the nearest neighbour distance. Other distances, d, are quoted as their squares and n, the number at that distance, is listed.

Table 1. Number of neighbours and distances apart

	NaCl		CsCl	
	d^2	n	d^2	n
unlike	1	6	1	8
	3	8	11/3	12
	5	24		
like	2	12	4/3	6
	4	6	8/3	12
	6	24	4	8
			16/3	6

This Table shows how the two lattices differ. The number of nearest neighbours alone does not determine the relative stability of the lattices, as it did for the inert gas solids, because neighbours of many orders contribute substantially to the electrostatic effects.

It is usual to index the planes of the lattice by using the vector which is the normal to the plane from the origin. Thus (100) denotes a plane normal to the x-axis, the yz plane. More exactly, this index refers to the set of planes parallel to this plane which, together, pass through all the lattice points. Similarly (110) is a plane normal to the vector bisecting the angle between the x and y axes and (111) is normal to the space diagonal.

2-3 Electrostatic forces

The calculation of the potential of an ion due to all its neighbours is a complex task. The terms in the summation decrease very slowly while the numbers at a given distance tend to increase so that the sum is only conditionally convergent. The efficient calculation of these sums requires more elaborate techniques than are appropriate to describe here. A brief account is given in Appendix 1. The result of these calculations can be given easily. The sum of the potentials over these lattices is expressed as a Madelung constant divided by the nearest neighbour distance. These constants are listed in Table 2.

Table 2. Madelung constants			
NaCl	1.747558	CsCl	1.762670

This shows that the CsCl lattice has more electrostatic cohesive energy than the NaCl lattice but the difference is not great. Because of the number of nearest neighbours the CsCl lattice will have more repulsive energy.

2-4 Ionic radii

The simplest model of the ionic crystal treats the ions as rigid spheres with fixed radii. The sum of the two unlike radii should be the distance between two nearest neighbour ions in the crystal if these are touching. This simple result of the model gives a set of relations between the ionic radii and the experimental interionic distances in the crystals which are obeyed to a good approximation. The determination of individual radii is more difficult and cannot be completed from interionic distances alone. It has been discussed fully by Tosi and Fumi [3] using a more elaborate comparison with experiment and they recommend the following values. These values are smaller by about 0.2Å than the traditional ones for the anions but are correspondingly larger for the cations.

Table 3. Radii of Ions (in Å)

cation	Li+	Na+	K+	Rb+	Cs+
	0.90	1.21	1.51	1.65	1.80
anion	F⁻	Cl⁻	Br⁻	I⁻	
	1.19	1.65	1.80	2.01	

With these radii it is now possible to calculate the electrostatic energy of a crystal. There is one complication. It is readily assumed that the ions of opposite charge are touching and that this fixes the dimensions of the crystal but, when the cation is small compared with the anion, it is the anions themselves which touch and determine the size of the cube. The criterion for this, in the NaCl structure, is that the ratio of the radii, $R+/R-$, is less than $\sqrt{2}-1=0.414$. These new radii show that this happens less frequently than had been assumed in the past. For the CsCl lattice this ratio should be less than $\sqrt{3}-1=0.732$. Thus this lattice, which is favoured by the electrostatics, is possible only when the ions are nearly of the same size.

This simple model, as for the corresponding rigid-sphere model for the inert-gas crystals, is too simple to allow many properties to be calculated. It takes no account of the nature of the repulsive force and ignores any contraction of the crystal due to the strong attractions.

2-5 Scaling the energy

If the repulsive force between two ions of opposite charge is taken as the inverse nth power, the energy of a univalent ion pair will be (in atomic units, see Appendix 8):

$$u = \frac{a}{r^n} - \frac{1}{r}$$

In the NaCl lattice the cohesive energy per ion pair will then be:

$$U = \frac{6a}{r^n} - \frac{\gamma}{r}$$

where γ is the Madelung constant and each ion has 6 nearest neighbours touching it. This can be rewritten as:

$$U = \frac{\gamma}{d}\left(\frac{1}{n}\left(\frac{d}{r}\right)^n - \frac{d}{r}\right) = \frac{1}{d} f\left(\frac{r}{d}\right)$$

where d is the distance at which the energy, U, is a minimum. This shows that the energy, for any pair of ions adopting this crystal structure, can now be scaled using the equilibrium distance d as the unit of distance and 1/d as the unit of energy. A more extensive discussion of scaling in the theory of ionic systems is given by Blander [4].

Table 4. Interion distances (d) and melting points (T_m) for alkali halides

	T_m(K)	d (Å)	T_md/10
LiF	1121	2.01	225
LiCl	887	2.57	227
LiBr	823	2.75	227
LiI	718	3.02	221
NaF	1265	2.31	292
NaCl	1074	2.81	302
NaBr	1023	2.98	304
NaI	933	3.23	301
KF	1129	2.67	302
KCl	1045	3.14	328
KBr	1013	3.29	334
KI	958	3.53	339
RbF	1048	2.82	296
RbCl*	988	3.29	326
RbBr*	953	3.43	327
RbI*	913	3.66	334
CsF	955	3.01	288
CsCl*	918	3.47	318
CsBr*	909	3.62	329
CsI*	894	3.83	342

* In the CsCl lattice

The scaled energy becomes independent of the particular ions. We note that this depends on the value of n being fixed. For the CsCl lattice the calculation is similar but the function f will be slightly different since the Madelung constant is different. As in Chapter 1, the scaling of the energy means that any properties that depend only on the energy can now be scaled and the scaled properties will be constants for all the ions that have this crystal form. Table 4 shows the interionic distances and the melting points of the alkali halides. This table shows that the scaled melting point is close to being constant despite possible differences in the value of n. Even the fact that some crystallize in the CsCl structure has little effect on the result.

Another property which can be related by scaling [4] is the surface tension of the liquid measured at the melting point $\sigma(T_m)$. This has the dimensions of an energy per unit surface area and so is scaled by multiplying by d^3. The result of this is shown in Table 5.

Table 5. Surface tension at the melting point

	σ	d	$\sigma d^3/10$
LiF	251	2.01	204
LiCl	138	2.57	234
NaF	202	2.31	249
NaCl	116	2.81	257
NaBr	99	2.98	262
NaI	88	3.23	297
KF	142	2.67	270
KCl	99	3.14	306
KBr	89	3.29	312
KI	79	3.53	348
RbF	131	2.82	294
RbCl*	99	3.29	352
RbBr*	91	3.43	367
RbI*	83	3.66	407
CsF	107	3.01	292
CsCl*	90	3.47	376
CsBr*	85	3.62	403
CsI*	75	3.83	421

This table shows that the scaled surface tension is more sensitive to the simplicity of the model than is the melting temperature itself. In particular, there is some evidence that the CsCl structures have a larger value for the scaled surface tension and this may reflect their larger value of γ.

As in Chapter 1, the scaling method gives interesting results with the minimum of effort so long as the property is determined by the energy formula alone. If the ionic mass, the external pressure or temperature become involved in the property then the method is no longer relevant.

2-6 The Pauling rules for ionic crystals

When an ionic crystal contains three or more distinct ionic species the form of its lattice becomes more difficult to predict. This is especially so when some of the ions are multiple ions. Often a cation has a multiple charge, and so has a very small radius, whereas the anion is singly or doubly charged and has a large radius This suggests a difference in their roles in determining the structure. Four general rules about the stability of such structures have been given by Pauling [5].

The first rule is:
> *A coordinated polyhedron of anions is formed about each cation,*
> *the cation-anion distance being determined by the radius sum*
> *and the coordination number of the cation by the radius ratio.*

The number of anions can be eight, forming a cube, or six, forming an octahedron. As mentioned above, there is a geometrical effect which arises if the ratio of the radii is so small that the anions in the cube begin to touch. In these structures the consequence of too small a ratio is that a structure with a smaller coordination number becomes more stable. This critical value between the cube and the octahedron is given by

$R+/R- < \sqrt{3}-1 = 0.732.$

There is a similar transition from octahedral to tetrahedral coordination when the ratio falls below $\sqrt{2}-1 = 0.414$.

The charges on the ions have an obvious importance since they govern the electrical attractions. The second Pauling rule is:

In a stable ionic structure the valence of each anion, with changed sign, is exactly or nearly equal to the sum of the strengths of the electrostatic bonds to it from the adjacent cations.

The strength of the electrostatic bond is defined as the ratio of the number of charges to the coordination number. This concept of strength is very similar to the division of the ionic charges used in Appendix 1 in the definition of supercells. Thus if the anion charge is $-\zeta$, in units of the proton charge, then it satisfies, at least approximately,

$$\zeta = \sum_i \frac{z_i}{n_i}$$

where z_i is the charge of each touching cation and n_i is the number of anions around that cation. This rule ensures that each anion will be placed in the strongly attractive field due to the cations. The electrostatic forces favour surrounding each charge by nearest neighbours of opposite charge to neutralise it, while allowing for the division of these charges when they are shared. Thus, for example, the cation Ti^{4+} is surrounded by O^{2-} ions in certain crystals. Their radius ratio is 0.49 which is below the cube limit but higher than the limit where the octahedron ceases to be preferred so its coordination number will be six. The strength of the Ti^{4+} will be 2/3. In $BaTiO_3$ each O^{2-} has two of these cations and also four Ba^{2+} with a strength of 1/6 so the doubly-charged anion will be balanced by the sum of these strengths.

The third Pauling rule concerns the sharing of corners, edges or faces between polyhedra. It states that:

The presence of shared edges and especially of shared faces in a coordinated structure decreases its stability; this effect is large for cations with large valence and small coordination number.

The reason for this rule is electrostatic repulsion rather than electrostatic attraction. Shared elements bring the enclosed cations close together and so increase their repulsion. The geometry shows that the smaller the coordination number the more pronounced the effect will be. Thus, in some crystals such as silica, the Si^{4+} ion is surrounded by O^{2-} ions with a radius ratio of 0.37. This low ratio predicts tetrahedral coordination. No crystal is known in which these tetrahedra share a face or even an edge.

The final rule is closely related to the third one. It says:

In a crystal containing different cations those with large valence and small coordination number tend not to share polyhedron elements with each other.

This is another way of making sure that the cations with large charges are kept as far apart as possible.

By applying these rules, as Pauling has demonstrated, the structures of many particular ionic crystals can be understood and predicted. An extensive discussion has been given by Clark [6].

2-7   Ferroelectrics

As a consequence of the long-range of the Coulomb interaction it is to be expected that the ionic crystals will be rich in properties which involve cooperative effects. One of these is ferroelectricity. This arises when dipoles in one unit cell help to polarize the next cell and induce parallel dipoles there. The combined effect may be to give the crystal a macroscopic dipole moment. A simplified account of this cooperative effect is given in Chapter 7.

One class of structures which can show ferroelectric behaviour is the $BaTiO_3$ group. The basic structure of these is the perovskite structure shown in Figure 3. As the Pauling rules indicate, each Ti^{4+} has an octahedron of O^{2-} around it and it is situated at the body centre of a simple cubic lattice which has the Ba^{2+} at its vertices. The octahedra share vertices. The rather larger Ba cations have twelve O anions touching them (radius ratio = 0.96) and these are at the face centres of the Ba lattice. This is the high temperature form of the solid (above

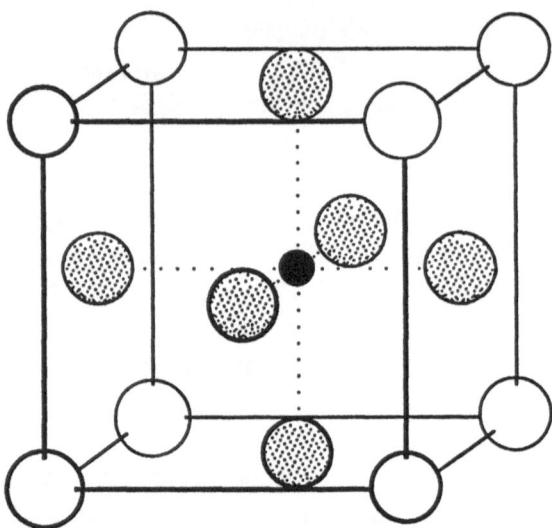

Fig. 3 Perovskite structure

120°C) and it is not ferroelectric. As the temperature is reduced the material undergoes transitions to three other crystal structures which do show ferroelectricity. Between 120°C and 5°C the crystal is tetragonal with the polarization directed along one of the axes of the original cube. Between 5°C and -80°C it is orthorhombic with a distortion, and polarization, in the direction of a face diagonal of the cube while, below -80°C it becomes rhombohedral with distortion along the cube space diagonal. These can be understood qualitatively in terms of the movement of the Ti^{4+}. It was pointed out (Megaw, [7]) that the Ba ions are larger than can be accommodated within the coordinated O octahedra so that the latter must expand. This means that the Ti ion becomes a little loose inside its octahedron so that at lower temperatures it settles close to one, two or three of its surrounding anions. These unsymmetrical displacements result in dipoles in different directions and these are the origin of the cooperative electrical effect. In the lowest temperature form, for example, the ion is at a face of the octahedron touching three anions and gives a dipole in the direction of the cube diagonal. In practice the movement of the anions is also important and the geometry of the whole lattice structure is slightly disturbed in each of the ferroelectric phases. The cooperative electrostatic effect must then be related to the elastic forces arising from this distortion. Further comments are found in Chapter 7.

A phenomenological theory of this effect has been given by Devonshire [8] and modified by Slater [9]. They expand the free energy in terms of the components of the polarization of the unit cell with terms of quadratic, quartic and sextic degree. It is essential in these calculations to allow for the high polarizability of the O^{2-} ions. The results are in good agreement with the experimental observations.

2-8 Superconductors

Recently, there has been a major development in the study of superconductivity due to the discovery of ceramic materials which exhibit this effect at much higher temperatures than previously thought possible. The crystal which has been most discussed is that of $YBa_2Cu_3O_{7-\varepsilon}$. This has a structure of perovskite type (a full description is given by Robinson [10]). The Cu cations form a simple cubic lattice. Three cubes, in a vertical direction form the unit cell. The top and bottom cubes have Ba cations at their body centres while the centre cube has Y or some other lanthanide. In the horizontal layers through the Ba, the O anions are situated at the mid points of the vertical edges of the cubes. On the other hand, the horizontal layer through the heavy ion has no O anions. Half the O anions are missing from the edges of the top and bottom faces. These missing ions mean that the central vertical plane in one direction has many fewer O anions than the other central vertical plane. It also means that the O ions in the planes above and below the Y ion will distort slightly from the cube edges to fill the vacant spaces. The structure becomes orthorhombic rather than cubic. The crystal is superconducting up to 98 K.

The structure has some significant departures from the perovskite crystal symmetry since no O octahedra remain. There is also a small deficiency of O from the stochiometric ratio (ε can be as much as 0.19). This deficiency is probably necessary to produce the carriers of the current. Some of the distortion of geometry is due to partial covalent character (see 2.13 below) in the bonding between the Cu and O which tends to form them into linear horizontal chains of CuO_2 molecules though there remains some connection between the chains in the same horizontal layer. The evidence from experiment is that the superconducting currents are primarily carried in the top and bottom horizontal layers through these CuO_2 chains.

The theory of this effect is not yet established. It will have to explain the source of the current carriers as well as the nature of the interaction which permits them to operate at such temperatures. The fact that these crystals allow some planar distortions but are very resistant to the formation of point defects must be relevant to the absence of scattering of the charge carriers. The experimental fact that the carriers have zero spin argues that there is some coupling mechanism between electrons of opposite spin as there is in the BCS theory of metallic superconduction. Several theories appeal to the covalent-ionic character of the CuO_2 chains to suggest a conduction mechanism through resonating bonds.

2-9 Surface energy

When an ionic crystal is split into two by cleavage at a plane the energy required may vary considerably, depending on the direction of the plane. Some planes would leave one surface with ions entirely of one sign and the other surface with ions of the opposite sign. These energies would be very large. On the other hand, if the separating planes each contain balanced numbers of ions the energies will be much smaller. Thus there is a selection rule which forbids some cleavage planes. For the NaCl lattice the planes (111) are not allowed.

The energy needed to create a surface has two components. The energy to separate the ions in the two halves can be calculated from the interionic forces assuming that each retains perfect crystal geometry. The new surface ions are then no longer in equilibrium so that there will be a relaxation in their positions which will extend to lower layers. This calculation depends on the polarizability of the ions. In the regular NaCl lattice each ion is at a point where the electric field vanishes so that the polarizability of the ion does not contribute to the cohesive energy. When the lattice geometry is disturbed the ions may be subject to considerable electric fields and their polarization response will be important. This is especially so for the anions. This part of the energy will not scale in the same way as the first part because it introduces the polarizability as an extra variable. The calculated energies (Benson, [11]) for two cleavages of some alkali halide crystals are given in Table 6.

Table 6. Specific surface energies for alkali halides

	plane	10^{-6}J/cm^2
LiF	(100)	14.2
	(110)	56.8
LiCl	(100)	10.7
	(110)	34.0
NaF	(100)	21.6
	(110)	55.5
NaCl	(100)	15.8
	(110)	35.4
KF	(100)	18.4
	(110)	42.3
KCl	(100)	14.1
	(110)	29.8

This table shows that the (100) planes are the preferred cleavage planes. This leads to the cubic forms so often shown by the macroscopic crystals.

2-10 Defects

No crystal is perfect. The process of crystallization rarely leads to complete regularity over a large volume. In the ionic crystals there are dislocations where, over a line or a plane, the ions are out of position. In this section we are concerned with more localized faults which can arise through thermal disturbances. These are point defects. There are two basic forms of defect. The Schottky defect consists of two ions, of opposite charge, removed from their positions leaving vacant lattice sites. The Frenkel defect consists of an ion moved from its site to an interstitial position. There are two kinds of this defect depending on the charge of the ion which has been moved.

The energy required to create a vacant lattice site can be calculated from the interionic potential. The lattice is then left in a shape which is not in equilibrium so it will relax. The relayed effect of this on all the surrounding ions can also be estimated. Since the ions are no longer at neutral points of the electric field the effects due to the polarization of the ions must be included. Calculations differ on the models used to represent the distortion of the electron clouds. The anions will distort more in their valence shell than in their inner shells so that two different polarizabilities may have to be used to describe the response. The energies of formation of the different defects for some lattices (as quoted by Lidiard [12]) are given in Table 7.

Table 7. Formation energies of defects

	+ Frenkel	— Frenkel	Schottky
NaCl	2.88	4.60	1.79-2.34
NaBr	2.56	4.84	1.66-2.20
KCl	3.46	3.73	1.90-2.20
KBr	3.16	4.17	1.81-2.13

(The range of values for the Schottky defects results from slightly different theoretical models being used.) In all of these the Schottky defect has the lowest energy and so would be expected to be the most common of the crystal defects at room temperature. The fact that these energies do not have a constant ratio to the melting temperature is evidence for the importance of the relaxation effect since it is this part which does not scale.

The presence of defects in the crystal has a significant effect on various properties of the crystal. One of these is the mechanism of diffusion through the crystal. A vacancy moves when one of its neighbouring ions overcomes the barrier between them and fills the vacancy leaving its own site vacant. This gives rise to ionic conduction.

The height of the barrier is the activation energy which determines the temperature dependence of the diffusion and of the conduction. For a full discussion see Corish and Jacobs [13].

A similar discussion can be given for the entry of foreign ions or atoms into a crystal. Since these intruders rarely match in charge and size the ions which they displace there is relaxation around them. If the ionic charge is changed the extra electric field which this creates may have great significance for the electrical properties of the crystal. This "doping" will be discussed in a different context in 4-6.

2-11 Colour centres

When a free electron is attracted to an anion vacancy a new kind of defect arises. It is readily created in a salt by high energy radiation which produces colour in the transparent crystal so the name "colour centre", or F-centre, has been given to it. Around the vacant site the cations will attract the electron almost as if there was a charge at the centre of the vacancy. The electron will react rather like the electron in a H atom. There are two differences. When the electron is far from the vacant site the ions that lie between them will shield the field and this can be represented approximately by introducing a dielectric constant, ε, into the Coulomb attraction. Then, again, the electron, in its movement, is not free but is influenced by the periodic potential of the lattice. This can be partly represented by introducing an effective mass, m^*, which is related to the curvature of the conduction band of the structure at its lowest point. The result is a Schrödinger equation (we use atomic units, see Appendix 8) which is similar to the H-atom equation and can be solved exactly:

$$\left(-\frac{1}{2m^*}\nabla^2 - \frac{Z}{\varepsilon r}\right)\Psi = E\Psi$$

The energy eigenvalues (in Hartrees) are:

$$E_n = -\frac{m^*Z^2}{2\varepsilon^2 n^2}$$

where Z, the charge to neutralize the vacancy, is usually unity. From this model the excitation energy from the 1s to the 2p level (the origin of the colour) can be deduced and compared with experiment:

$$E_{2p} - E_{1s} = \frac{3m^*}{8\varepsilon^2}$$

The observations are in fair agreement with this but since the 1s orbital is much modified by the absence of a nucleus at the centre and the presence of a positive charge distributed over six cations its energy is poorly estimated.

A better model of the potential is obtained by averaging the actual Coulomb potentials of the neighbourhood ions (the Ewald field) over a spherical surface to give an effective radial potential for the same equation. The first two states can be solved by a numerical integration of the radial Schrödinger equation (Laughlin, [14]). Typical results are shown in Table 8.

Table 8. Energies of F-centres (in Hartrees)

	$-E_{1s}$	$-E_{2p}$
LiF	0.299	0.153
NaF	0.273	0.154
LiCl	0.254	0.153
NaCl	0.239	0.151
KCl	0.220	0.148
RbBr	0.206	0.144

The table shows that the upper state is insensitive to the crystal potential but the ground state has a small variation from crystal to crystal.

The F-centre is the first of a series of colour centres which contribute to these effects in a solid. When two anion vacancies are adjacent to one another and together trap one electron this is an M-centre and the model of a hydrogen molecular ion is appropriate. By scaling this model using ε and m^*, estimates of the electronic spectra are obtained. It is also possible to consider the possibility of trapping of two electrons between these two vacancies and use the hydrogen molecule as a model. Again, there can be three anion vacancies together, forming a R-centre, and its

properties will show a considerable dependence on the geometrical
configuration of these three.

2-12 Ionic melts

When a crystal melts, its components become free to diffuse throughout
the entire volume available to the liquid. This freedom is limited,
however, when the components are ions. This is shown by the radial
distribution functions for the liquid. An extensive review of the
literature on these distribution functions has been given by McGreevy
[15]. The main features of their forms are illustrated in Figure 4 which is
based on the results for NaCl which he quotes.

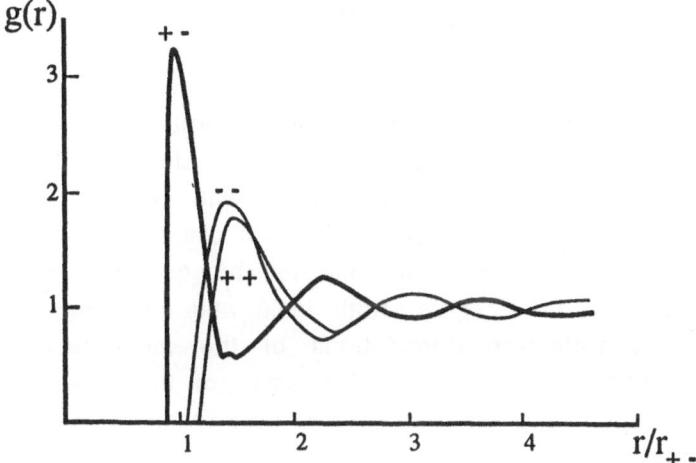

Fig. 4 Radial distribution functions for an ionic liquid

These results demonstrate clearly that, in the liquid, the cations are
still surrounded by anions and vice versa. The distances apart are no
longer fixed as they are in the solid so that the peaks in the functions are
broadened. They are also pushed further apart since the interionic
distance expands on melting. There is a clear indication that the
structure extends to several layers of ions since various peaks are

observed. Eventually, at larger separations, the structure loses its order and becomes random so that g becomes flat.

The implications of this observation are important. The short-range structure is maintained in the course of melting although the long-range structure of the crystal has been lost. It is clearly the strength of the electrostatic forces which produces this result. The general consequence goes beyond this since it follows that the nature of the liquid-solid transition depends on long-range order and not on short-range order.

The nature of the electronic energy levels in the ionic liquid has been studied by Parent *et al* [16] using a Green's function model (see Appendix 7).

2-13 Covalent and ionic character

The ionic character of the alkali halides is borne out in many of their properties. Other combinations of ions raise doubts about the extent to which the ionic character becomes mixed with covalent character in their solid state. The mixture would mean that electrons could pass more freely from one ion to the other and the net charges associated with each nucleus might not be integers. Both ionic and covalent bonding are approximate one-electron descriptions of the multi-electron reality. This means that an approximate discussion of the situation may be sufficient to explain the significance and occurrence of the two limiting types of bonding.

We start from a crude account of the covalent bond. This has an orbital directed from one centre towards the other and an answering orbital from the second centre pointing towards the first. The bond orbital constructed from these two will be the eigenvector of the corresponding matrix of the Fock Hamiltonian:

$$\begin{pmatrix} A & \beta \\ \beta & B \end{pmatrix}$$

where the matrix elements are defined as:

$A = \int \phi_A^* H_{eff} \phi_A d\tau$, $B = \int \phi_B^* H_{eff} \phi_B d\tau$, $\beta = \int \phi_A^* H_{eff} \phi_B d\tau$.

The orbital energies, the eigenvalues, are given by:

$\varepsilon = (A + B)/2 \pm \sqrt{(\beta^2 + (A-B)^2/4)}$.

For our purpose the important quantity is G, the gap between these. If we define the ionic character by C=A-B, which is the electronegativity difference, and the covalent character by H=2β, which is a measure of the bonding since it is related to the inverse of the hopping time for electrons, then the gap is given by

$G^2 = H^2 + C^2$.

The fraction of the bond which is ionic is defined to be

$f_i = C^2/G^2$

while the covalent fraction is

$f_c = H^2/G^2$.

These quantities have been used by Phillips [17] in his discussion of ionicity in crystals and its effects on crystal structure. He introduces values for the integrals in terms which can be found from experiment. Thus he defines:

$C = 1.5 \left(\dfrac{Z_A}{r_A} - \dfrac{Z_B}{r_B} \right) e^{-kR}$; $H = E_0 R^{-5/2}$.

where Z_A is the valence of A, r_A its covalent radius, R the distance between A and B, E_0 is constant for a given isoelectronic sequence and k is related to a Fermi gas of electrons with the same density.

Using his definition Phillips has calculated the values of C and H for many crystals as a function of crystal form. Table 4 lists a selection from his results. This shows clearly the correlation of the typical NaCl structure (denoted by S) with the crystals having $1 > f_i > 0.78$. The wurtzite structure (denoted by W) is found when $0.79 > f_i > 0.56$ and the ZnS structure (denoted by Z) when the value lies in the range $0.59 > f_i > 0.1$. The diamond structure (denoted by D) is found when $f_i = 0$.

It is interesting to compare this definition of ionicity with the one due to Pauling [5] which is better-known to chemists. Table 7 gives some selected values of both together with their crystal forms. This Table shows how the forms vary from covalent to ionic as this ionicity variable increases but the edges of the different regions do have some overlap. The Pauling values correlate fairly well with the Phillips values but do not discriminate so effectively when compared with the crystal forms.

Table 7. Ionicity of crystals using Phillips and Pauling definitions

	Form	Phillips	Pauling
CC	D	0	0
BN	Z	0.256	0.42
SiC	Z	0.177	0.11
AlP	Z	0.307	0.25
CuF	Z	0.766	0.92
ZnS	Z	0.623	0.59
AlSb	Z	0.426	0.26
GaAs	Z	0.310	0.26
GaSb	Z	0.261	0.26
InSb	Z	0.321	0.25
BeO	W	0.602	0.81
AlN	W	0.449	0.56
ZnO	W	0.616	0.80
MgSe	W	0.790	0.65
LiF	S	0.915	0.98
NaF	S	0.946	0.98
NaCl	S	0.935	0.94
CaO	S	0.913	0.97
CdO	S	0.785	0.85
CaS	S	0.902	0.81
KCl	S	0.953	0.95
KBr	S	0.952	0.91
RbBr	S	0.957	0.94

References

[1] Born M and Huang K 1954 Dynamical theory of Crystal Lattices, Oxford
[2] Löwdin P O 1948 A Theoretical Investigation into some Properties of Ionic Crystals, Almqvist and Wiksell Uppsala
[3] Tosi M P and Fumi F G 1964 J Phys Chem Solids 25: 53
[4] Blander M 1967 Adv. Chem.Phys. 11: 83.

[5] Pauling L 1945 Nature of the Chemical Bond, Cornell New York; 1929 J Amer Chem Soc 51: 1010

[6] Clark G M 1972 The Structure of Non-molecular Solids, Applied Science Publishers London

[7] Megaw H D 1946 Trans Faraday Soc 42A: 224

[8] Devonshire A F 1949 Phil Mag. 40: 1040.

[9] Slater J C 1950 Phys Rev 78: 748.

[10] Robinson A L 1987 Science 236:1063

[11] Benson G C 1963 J Chem Phys 39: 302

[12] Lidiard A B in : March N H 1974 Orbital Theories of Molecules and Solids, Oxford

[13] Corish J and Jacobs P W M 1973 Surface Defect Props. Solids 2: 160

[14] Laughlin C 1965 Solid State Comm. 3: 55

[15] McGreevy R L 1987 Solid State Physics 40: 247

[16] Parent L G, Davison S G and Ueba H 1980 J Electroanal Chem 113: 51

[17] Phillips J C 1970 Rev Mod Phys 42: 317

Chapter 3   Molecular Crystals

3-1   Introduction

In Chapter 1 we have considered the structure of solids whose atoms interact through van der Waals and repulsive forces alone and, in Chapter 2, the complication of introducing strong electrostatic forces has been added. In this Chapter we return to the simpler situation and consider the crystals of molecules which carry no net charge and have relatively small, or zero, dipole moments. Their binding is largely determined by the packing together of the molecules but some electrostatic effects may enter to complicate the discussion.

3-2 Forces between molecules

The simplest model of a molecule in a crystal gives each atom a radius and a spherical shape. The spheres for the atoms in neighbouring molecules are not allowed to overlap (although the spheres of atoms in the same molecule do so). These non-bonding radii can be estimated from observations on the closest distances of approach of their atoms when molecules are nearest neighbours in crystals. This implies some averaging since the radii do vary slightly from molecule to molecule. One survey of the crystal data suggests that the non-bonding radii (in Å) of the most common atoms should be:

    H   1.17
    C   1.80
    O   1.52
    N   1.58

By using these radii to construct the steric envelopes of molecules the fitting together of molecules in crystals can be investigated.

The crystal structure of many organic molecules can be understood as a close packing of the molecule into the unit cell. This idea has been used extensively by Kitaigorodsky [1] who measures the closeness by the packing coefficient K, defined as the ratio of the volume of the molecule to the volume available to it. The molecular volume is estimated as the sum of the volumes of the steric spheres, after allowing for their overlapping. The molecular geometry determines the centres of the spheres. In the crystal, the volume available is the volume of unit cell divided by the number of molecules it contains. For a large number of molecular crystals K has a value in the range:

$0.68 < K < 0.9$.

He also points out that as the temperature rises the solid will expand, due to thermal expansion, and this coefficient will fall to about 0.68 when melting will begin. As the crystal melts the ratio drops to about 0.58 and further heat makes the liquid gaseous at about 0.5.

We have seen, in Chapter 1, that the next stage in modelling crystal interactions is to introduce explicit expressions for the forces between molecules. In particular the interactions between molecules with high symmetry around a centre can be represented by dispersion and overlap forces between these centres. Molecules with little or no symmetry must be represented by forces which mimic the shape of the molecules more accurately than this. One way to do so is to use dispersion and overlap forces between each atom of the one and each atom of the other. This effectively prevents any atom of one approaching any atom of the other closer than their van der Waals radii permit, as in the steric spheres model, while still allowing for some attraction between more distant atoms.

Molecules, in general, have dipole moments and quadrupole moments. The electrical forces between molecules are not entirely negligible in the solid state since these moments may be lined up and so give cooperative effects. Nevertheless the magnitude of the electrical terms is usually small in comparison with the other terms so that the structure is largely determined by considerations of packing.

A theoretical analysis of the crystal structures of hydrocarbon molecules has been made by Kitaigorodsky [1] and Williams [2] (for a useful review see Ramdas and Thomas [3]). As a result, there are empirical values of the constants describing the forces between these molecules. The assumption is that the dispersion energy is an inverse sixth power with coefficient A and the repulsion an exponential with exponent α and coefficient B. The values of these are shown in Table 1.

Table 1. Constants for the atom-atom interaction of hydrocarbons

	A	B	α
C...C	2376.5	349908	3.60
C...H	523.0	36677	3.67
H...H	114.2	11104	3.74

(A in kJ/mole $\mathring{A}^6$, B in kJ/mole, α in 1/$\mathring{A}$)

By using these values as a foundation, the interaction constants involving other atoms can be evaluated empirically from the cohesive energies. Using these inter-atomic constants good estimates can be made of the interaction energies of molecules as functions of their distance apart and their orientation. The possible equilibrium configurations are then found by minimizing. This "molecular modelling" has been of great importance in understanding the structure of crystals and in contributing to the study of molecular interactions in liquids.

Much attention has been given to the representation of the electrical forces between molecules but this does not have an agreed solution. The simple idea of using the calculated Mulliken charges on each atom has several deficiencies. If the basis set is not "balanced", and there is no simple criterion for this, these charges can be quite unrealistic. Nor can any system of charges on nuclei alone represent the localized dipoles, such as those on the O atom, that give rise to hydrogen bonding. Both of these faults are remedied by the use of optimized population analysis [4]. This optimizes the representation of the charge density by local distributions on each atom and leads to atomic charges and dipoles which are much less dependent on the limitations of the basis set. Some authors (Stone [5], Bonaccorsi *et al* [6]) have preferred to use the first three or four moments on the various nuclei, or on the mid-point of a bond, to represent the charge density. This leads to an accurate expression for the molecular electrostatic interactions at large distances. For interactions between molecules which may come close, these expressions which use point charges and point dipoles do not include any penetration term and so are in serious error [7]. To avoid this, the use of some point charges and a few diffuse Gaussians with small exponents, to represent the outer part of the electron density, has been suggested (Tsujinaga and Hall [8]). This model also allows the polarization of the charges by an imposed field to be calculated by changing the positions of the diffuse functions in proportion to the field. When the electrostatic forces between molecules are strong, these polarization effects cannot be ignored. Models of the electrostatic effects can now be made as accurate as the subsequent calculation demands.

3-3 Crystal structure

Many flat aromatic molecules crystallize in the same crystal form. The explanation for this can be found using the idea of close packing. We consider first the packing in the plane. It is easily seen that molecules of arbitrary shape can be placed so that each is in contact with six neighbours (see the boots in Figure 1). First, one line of molecules is set up with each touching its neighbour closely. Then, the next line is exactly similar but has all its molecules rotated through π (i.e. C_2). By adjusting the angles of the molecules to the lines these two lines can be made so that each molecule in one line touches two in the next line. The pattern is then repeated with these lines alternating. Thus, when the other lines are added, each molecule is touching six others. The unit cell is a parallelogram and there are C_2 rotations about each corner, about the middle of both edges and about the cell centre. Two of these molecular planes will not be close-packed vertically, if placed directly above one another, because of the three-dimensional shapes of individual molecules.

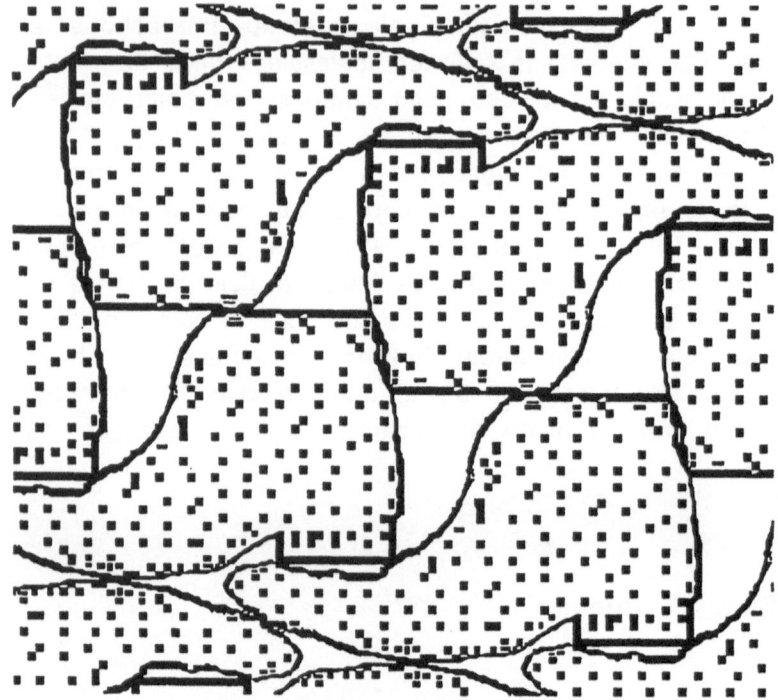

Fig.1 Packing in the plane

If the planes are slightly displaced and the molecules slightly rotated (while maintaining their touching) then the protuberances of one can more nearly fit into the hollows of the other. Thus each molecule close-packed in the plane becomes a stack of molecules, which are slightly tilted, in the solid.

There is also a close-packing of molecules which uses a glide-plane instead of a rotation to form the second line from the first (see Figure 2). The result also allows each molecule to touch six nearest neighbours. Its unit cell is rectangular and the parallel glide planes are along one edge and through the centre. This packing is used in the vertical packing which follows the horizontal packing described above.

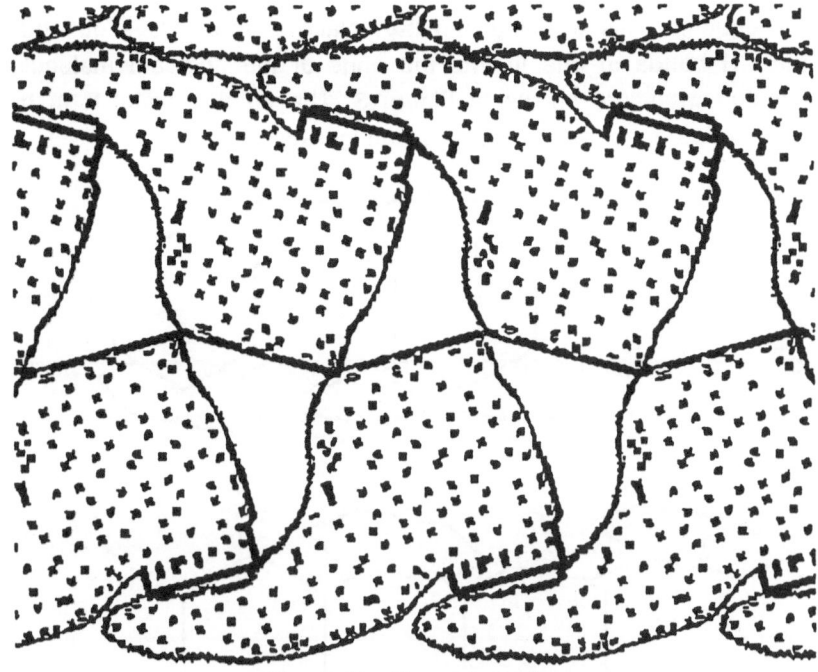

Fig. 2 Packing of stacks

In a plane which is normal to the first horizontal plane the stacks will appear as shapes which are close packed in this form. As can be seen, the stack in the centre of the cell will be composed of molecules tilted in the opposite way to those at each corner. Note the characteristic "herring-bone" shape of this packing due to its basic glide plane symmetry.

The significance of this structure, which forms the monoclinic space group $P2_1/a$ $\left(\begin{smallmatrix} 5 \\ 2h \end{smallmatrix}\right)$, is that close packing is achieved in three planes with 12 nearest neighbours for each molecule. Many polycyclic molecules crystallize in this form but some, for example pyrene, first form loose dimers and these are the units that take on this close packed structure.

The structure of the anthracene crystal has been studied extensively by Craig *et al* [9]. Its crystal form has the symmetry of the space group just described. The unit cell has basis vectors of lengths, 8.56Å, 6.04Å and 11.16Å. Its angles are 90°, 124.7° and 90° and the packing coefficient is K=0.718. These authors show that the molecules in alternate stacks are wedged together. This wedging can take place in three ways depending on how the CH bonds of one fit into the rings of the adjacent molecule. The two common ways are shown in Figure 3. The stable crystal form has the molecules related as in the bottom configuration.

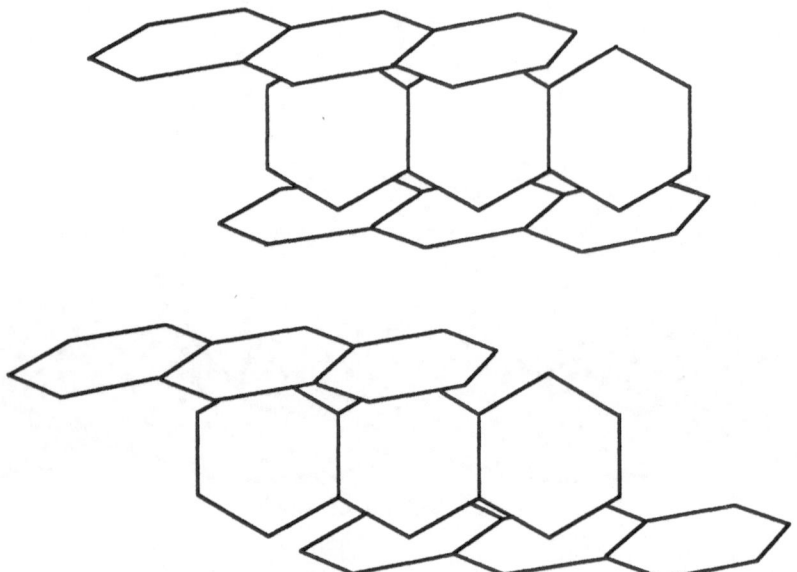

Fig. 3 Contact of two anthracenes, 2:3 above and 2:2 below

The upper configuration can be induced by the application of external pressure and may occur as a defect within the stable form. The calculations show that the energy difference between these forms is small.

Another interesting example of the use of molecular potentials to investigate structure is in the structures of the 1,5 dichloro anthracene crystal. This exists in two forms, monoclinic and triclinic. The monoclinic form allows two adjacent molecules to be lined up with their substituents directly below one another. When u.v. light shines on the crystal, dimers are formed and these have the structure head-to-head. On the other hand the triclinic structure gives 80% head-to-head and 20% head-to-tail. A calculation, based on a model consisting of about 20 molecules, (Ramdas *et al* [10]) suggests that disorders involving the reversal of some molecules occur more readily in this form. These would explain the occasional head-to-tail forms of the dimer derived from this crystal.

In some molecular crystals there is a plastic phase. This occurs close to the melting point when the molecules begin to rotate within the crystal. If the molecule is near spherical this rotation will not disturb the environment sufficiently to cause melting. In some examples the molecules act together in pairs for this purpose. Thus some diatomics, such as O_2, form dimers whose centres lie on a fcc lattice and may rotate.

3-4 The paraffin crystals

The crystal structure of the normal paraffins C_nH_{2n+2} raises some interesting problems. The melting points (reported by Schaerer *et al* [11]) can be plotted as a function of n, the number of C atoms. This is shown in Figure 4. This shows clearly that there are two trends. The temperatures of melting of the molecules with n an even number and $9<n<23$ follows a different curve from that followed by those with an odd number.

The explanation for this is found in the structure of the crystals. These are long-chain flexible molecules whose C atoms form a planar zigzag in the crystal. They tend to stack together in layers with their molecular axes parallel but tilted to the layer. The even-numbered molecules form triclinic or monoclinic crystals while the odd-numbered ones are orthorhombic. Thus the differences in melting point can be traced back to their different crystal structures.

The crystal structures should be considered in two stages. The molecules are first stacked together to form a layer and then the layers are brought together. The packing of the molecules is governed by the fitting of the CH bonds of one into the spaces between the two corresponding CH bonds of its nearest neighbour. For the even-numbered

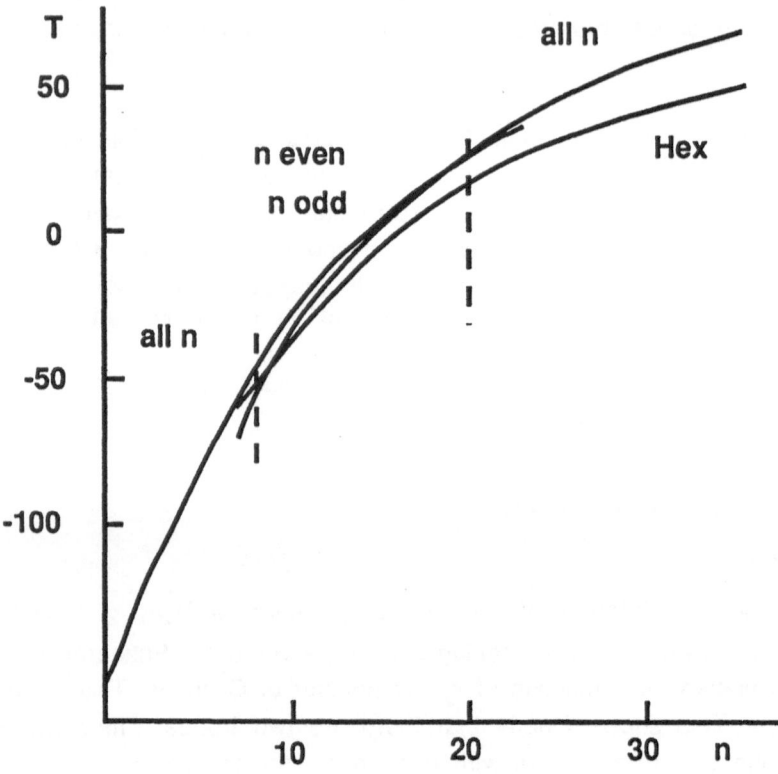

Fig.4 Melting points of normal paraffins

molecules this is done by translating one along its own axis relative to its neighbour so that the C atoms of one lie between the corresponding C atoms of its neighbour. The ends of the molecules will then lie on a plane which is oblique to the molecular axis. In this position it is possible for each molecule to be touching six nearest neighbours. This is the triclinic form. When there are more than 24 C atoms the monoclinic form is preferred. This has the molecules with their ends and their C atoms level but alternate molecules are twisted around their axes so that the CH bonds avoid one another. The odd-numbered molecules form a layer rather similar to this with the ends level. The difference comes in the packing of the layers. The layers cannot be superimposed exactly on one another since the projections will then interfere. Some staggering is required. In the monoclinic form the layers are staggered in the same direction from layer to layer so that the unit cell has its c axis at an angle to the molecular axis. In the orthorhombic form the layers alternate in stagger so that a layer will lie exactly above its next nearest layer below. The stability of the layers of these molecules and their ability to slip over the adjacent layers in the liquid state are the important properties which enable them to be used as lubricants. They form a protective layer on a metal surface which inhibits friction.

There is also a hexagonal phase which the odd-numbered molecules display for n=9 upwards. This has the molecular axes parallel and aligned so that the cross-section shows close packing with six molecules touching. The orientation of each molecule around its molecular axis is random. This is like the plastic phase which we mentioned in Chapter 1. The even-numbered molecules can also have this phase when the number of C atoms is 22 or more.

It is natural to ask whether other long chain molecules behave in a similar fashion. The related carboxylic acids can be considered. As usual, these will dimerize with the acidic groups hydrogen bonded. This doubles the length of the chain! These dimers will behave in a similar way to the paraffins. Now, if two double bonds are introduced into the molecules at the acidic end (i.e. R-CH=CH-CH=CH-COOH) the result is to make the molecule much less flexible in its central region. It can then form a nematic liquid crystal. In this phase the ends of the molecules are at random although their axes remain parallel and untilted. The lateral packing is hexagonal so that rotation about the molecular axis is also possible. The molecules are thus at random in two different respects.

3-5 Glass

The structure of the silica crystal (SiO_2) can be predicted from the Pauling rules given in Chapter 2. Because of its small radius ratio the Si^{4+} is surrounded by four O^{2-} in a tetrahedral conformation. Each oxygen ion belongs to two tetrahedra so that its strength is 1 and there is a balance with the charge of four on the Si ion. Nevertheless the bonding is not entirely ionic. Each SiO has some covalent character. The form of the crystal is diamond-like (see Chapter 4) with the Si at the positions of the C atoms and the O bisecting the CC bonds. The Si-Si separation is large so the crystal is very open. For this reason crystallization tends to be slow. If the liquid is cooled quickly it will remain a supercooled liquid until it reaches a temperature T_g ($T_g < T_m$) where it begins to form a glass. For the same reason the solid can easily fit additional ions such as Na^+ into the voids. The effect of these ions is to reduce the melting point of the solid and make it easier to mould.

The structure of this glass is not regular in the sense of forming a lattice but it does exhibit various regularities. Each Si ion is still surrounded by four O ions but instead of forming six-membered rings, non-planar like c-hexane, rings of different sizes and conformations are formed. The nature of the solid has been illuminated by computer simulations of its structure (see the review by Wooten and Weaire [12]). These follow the rule that each ion must be surrounded by the correct number of nearest neighbours (perfect short range order) but that no long range order is imposed.

Similar structures are found for other amorphous materials. In particular amorphous silicon and germanium have the same emphasis on short-range order at the expense of long range order.

3-6 Molecular vibrations in crystals

A critical test of the accuracy of the molecular modelling of the forces between molecules is to predict the elastic constants of the molecules within the crystal. These are deduced from the second derivatives of the energies with respect to various displacements. From these the vibrational spectra of the crystal can be calculated and compared with the observed IR and Raman spectra. The general theory has been developed by Born and Huang, [13].

For the crystals of large organic molecules the first approximation is to consider the molecule itself as a rigid body and calculate the vibrations that it has against its environment. Its centre of mass may move in three independent directions and it can rotate around three independent axes. From these degrees of freedom are derived six vibration bands which are usually called "acoustic bands" since they include the motions induced by sound waves. This theory has been applied by Pawley [14]. In the second approximation the effect of coupling the motions of the molecules to the internal vibrations of the molecule are considered. It is found that these frequencies are changed from those of the free molecule but individual vibrations have to be considered separately.

3-7 Excitons

Molecular crystals show several different types of electronically excited state. We consider first two identical molecules. Each has a lowest excited state so the pair, to a first approximation, has degeneracy. This degeneracy is resolved by an interaction between the two states. The position can be explained by using the simple product wavefunctions for the pair:

$$\Psi_1 = M_1{}^* M_2 \, , \quad \Psi_2 = M_1 M_2{}^*$$

where M_1 and M_2 are the wavefunctions for the ground states of the two molecules and the star denotes the first excited state. These two wavefunctions will interact with a matrix element S so that the Hamiltonian with respect to these two alone is:

$$\begin{pmatrix} U & S \\ S & U \end{pmatrix}$$

where U is the monomolecular excitation energy. The eigenvalues of this are $(U \pm S)$ and the eigenstates are $\Psi_1 \pm \Psi_2$. This effect is known as the Davydov splitting.

When the excited state, $M_1{}^*$, is an allowed singlet with a vector transition moment t_1 from the ground state, then the interaction matrix element between Ψ_1 and Ψ_2 can be approximated, to good accuracy, by the expression:

$$S = \Big(R^2(t_1.t_2) - 3(t_1.R)(t_2.R) \Big)/R^5$$

where **R** is the vector distance between the molecules. This is exactly the expression for the energy of interaction of two permanent dipoles. The transition moments of the new states are obtained by vector addition or subtraction of the two transition moments and so depend on the orientation of the molecules.

In a crystal of identical molecules, with one molecule in each unit cell, the same argument can be applied. The new excited state, using the phase difference method of Appendix 2, is a plane wave with wave vector **k**:

$$\Psi = \sum_r e^{i\mathbf{k}.\mathbf{r}} \Psi_r$$

where each Ψ_r is a product of molecular wavefunctions with the rth in its excited state and the remainder in ground states and **r** is its position vector. This is the Frenkel exciton. The excited energy band is given by

$$W(k) = \sum_s H_{rs} e^{i\mathbf{k}.(\mathbf{r}-\mathbf{s})}$$

which is a function of **k** and involves a sum over dipoles in all the molecules with the phase factor $e^{i\mathbf{k}.\mathbf{r}}$. It is easier to represent this sum by considering the electric field at **r** due to a dipole at **s** which is

$$E(s) = \left(-t_s R^2 + 3(t_s.R)R \right)/R^5$$

The total field acting on the molecule at **r** is given by

$$E_{eff} = \sum_{s \neq r} E(s) e^{i\mathbf{k}.(\mathbf{s}-\mathbf{r})}$$

$$= \sum_{s \neq r} \left(-t_s R^2 e^{i\mathbf{k}.\mathbf{R}} + 3(t_s.R)R e^{i\mathbf{k}.\mathbf{R}} \right)/R^5$$

The energy band is now
$$W(k) = E - t_r.E_{eff}$$

The evaluation of such sums is complicated since the results can be dependent on the shape of the crystal. The position is exactly like that of dipole sums in electrostatics. The exciton states which are created by absorbing radiation from the ground state lie close to k=0 so the sum for this value of k is the important one. In the Lorenz-Lorentz treatment of dipolar interaction [15] the summation is divided into two parts by means of a concentric sphere around the molecule at **r**. It is easily shown that the sum of the dipolar terms on a cubic lattice inside this sphere vanishes. Most molecular crystals do not have this degree of symmetry so that there is an internal contribution, E_i. In the simplest example this could be the field from the nearest neighbours and the sphere would be the one which encloses them. In practice rather larger spheres are preferred so that the total becomes independent of the size of the sphere. The sum of the dipoles outside the sphere is found by replacing the finite sum by a continuous polarized medium and using macroscopic electrostatics. The polarization can be estimated as

$$P = te^{ik.r}/v$$

where v is the volume of unit cell. The contribution to the field due to this continuum with its spherical hole is given by

$$E + \frac{4\pi P}{3}$$

if k is small. From the electrostatic equations

$$\nabla.(E + 4\pi P) = 0, \quad \nabla \wedge E = 0$$

the field is determined as

$$E = -4\pi(P.k)k/(k.k)$$

so that the energy is

$$W = (U - t.E_i) + 4\pi(k.t)^2/vk^2 - 4\pi t^2/3v.$$

Thus, because of the second term, the energy is non-analytic at k=0. This is the origin of all the problems of the summation. In particular, if k is parallel to t, even at k=0, then

$$W_= = (U - t.E_i) + 8\pi t^2/3v$$

and, if k is perpendicular to t which is the situation when the exciton is created by light,

$$W_\perp = (U - t.E_i) - 4\pi t^2/3v.$$

When this theory is applied to crystals such as anthracene account has to taken of the fact that there are two molecules per unit cell and these are not parallel. If the excited states of the cell are found first by treating just two molecules then the wave solution can be applied to the lattice of cells. Thus the original single level is already split inside the cell and there will be two bands of exciton waves. Detailed calculations have been carried out by Craig et al [16] for anthracene. The result, for the intense transition around 2500 Å, is that the calculated split is 14,600 cm^{-1} as compared with the observed split of 14,000 cm^{-1}.

This exciton is free to move through the crystal. In practice it may be stopped by an impurity, a dislocation or a phonon. In particular, if the crystal contains an impurity which can absorb light within the exciton band but fluoresces at longer wavelengths the energy may be trapped and eventually released as fluorescence typical of the impurity. The presence of small amounts of tetracene in crystals of anthracene can be detected in this way.

There is also another kind of exciton which can occur in molecular crystals. This is the Wannier exciton which is formed when an electron is removed from one molecule and given to another. If the two molecules are nearest neighbours this is called a charge transfer exciton. Its energies can be estimated from the ionization potential, electron affinity and geometry of the two molecules. More generally the two charges may be delocalized and the model of the H-atom like impurity centre in Section 2-11 can be adapted to estimate the energies that are possible.

3-8 Crystal engineering

The term "crystal engineering" was first used by Schmidt [17] to emphasise the importance of the use of crystals to influence the behaviour of chemical reactions. Useful reviews have been given by Thomas *et al* [18] and in the book by Wright [19]. The crystal environment determines the orientation of neighbouring molecules which may then react. By arranging the environment appropriately, yields of the desired products may be considerably enhanced. The available techniques for this purpose include:

1. The use of different crystal forms
2. External modification e.g. inducing a phase change to a desired configuration by external pressure
3. Co-crystallization with inert molecules which may orient the reactants favourably
4. Use of H-bonds to stabilize a desired geometry; perhaps introducing substituents to achieve this
5. Introduction of bulky substituents to modify the crystal packing

These techniques aim to eliminate the steric factors and probability factors, which make reactions in gas or liquid so slow, by bringing the reactants together in the solid in the most favourable configuration. They can also reduce the activation energy by implanting any charged species in a dielectric medium which will modify the forces.

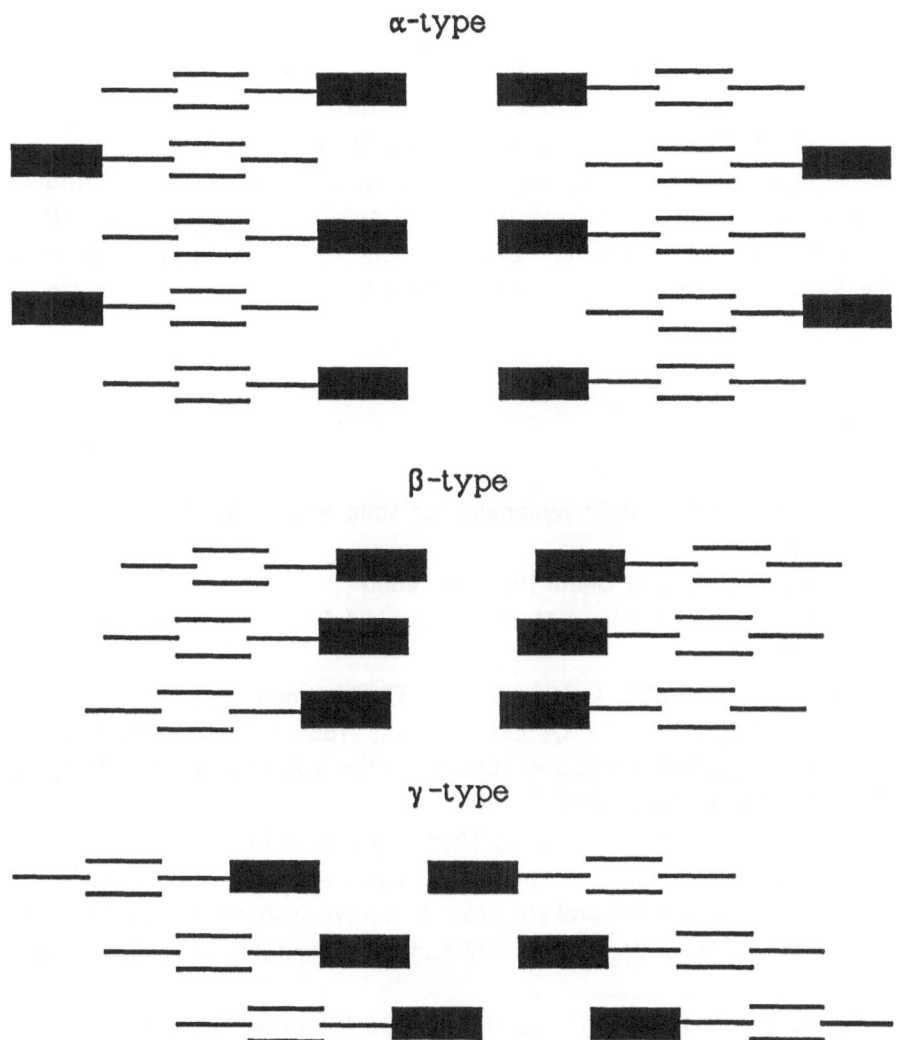

Fig. 5 Crystal forms for trans-cinnamic acid

The use of molecular modelling has become a vital tool in predicting the geometries of all these combinations. In practice the most common type of reaction to be studied in this connection is the photochemical reaction because the process is so readily under the control of the experimenter.

A good example is photodimerization. A well-known example of this is the dimer of trans-cinnamic acid. This substance crystallizes in three different forms depending on the method of preparation (Figure 5).

In the α phase the molecules are stacked up so that nearest neighbours are in opposite orientations. When the dimer is formed by irradiation these molecules link across their double bonds and form the centrosymmetric dimer. On the other hand, the β phase has the molecules stacked more nearly on top of one another so that irradiation produces the mirror-plane dimer. In the γ phase the stacks are such that the double bonds are too far apart to react and no dimer is formed. Thus, by selecting the mode of preparation of the crystal, the dimer of choice can be produced.

References

[1] Kitaigorodsky A J 1973 Molecular Crystals and Molecules, Academic Press New York
[2] Williams D E 1967 J Chem Phys 47: 4680
[3] Ramdas S and Thomas J M 1977 Chemical Physics of Solids and their Surfaces, 7: 31
[4] Smith C M and Hall G G 1987 Intern J Quant Chem 31: 685
[5] Stone A J 1981 Chem Phys Lett 83: 233; 1985 Mol Phys 56: 1047
[6] Bonaccorsi R, Scrocco E and Tomasi J 1976 J Am Chem Soc 98: 4049; 1977 J Am Chem Soc 99:4545
[7] Hall G G and Tsujinaga K 1986 Theor Chim Acta 69: 425
[8] Tsujinaga K and Hall G G 1986 Theor Chim Acta 70: 257
[9] Craig D P, Ogilvie J F and Reynolds P A 1976 JCS Faraday II 72: 1603
[10] Ramdas S, Jones W, Thomas J M and Desvergne J P 1978 Chem Phys Lett 57: 468
[11] Schaerer A A, Bayle G G and Mazee W M 1956 Rec Trav Chim Pays Bas 75: 2017
[12] Wooten F and Weaire D 1987 Solid State Physics 40: 2
[13] Born M and Huang K 1954 Dynamical Theory of Crystal Lattices, Oxford
[14] Pawley G S 1972 Phys Status Solidii 49b: 475
[15] Hall G G 1962 Proc Roy Soc A270: 285

[16] Craig D P and Walmsley S H 1968 Excitons in Molecular Crystals, Benjamin New York
[17] Schmidt G M J 1971 Pure Appl. Chem 27: 647
[18] Thomas J M, Morsi S E and Desvergne J P 1977 Adv. Phys Org Chem 15: 63
[19] Wright J D 1987 Molecular Crystals, Cambridge

Chapter 4   Valence Crystals

4-1 Introduction

In this Chapter we turn to a type of solid structure whose bonding is very different from those already discussed. A valence crystal is essentially one large molecule with valence forces alone acting to determine its structure.

The diamond crystal is the prototype of the class of crystals that now concern us. The concept of the tetrahedral valence of the C atom has been of immense importance throughout organic chemistry and in the diamond crystal it reaches its brilliant macroscopic consummation.

In order to lead into the subject in a simple way we start by considering the structures of some polymers which are equally determined by valence forces but are periodic in only one direction. These polymers are of importance for our study in their own right because they are examples of the newer materials now being synthesized which sometimes show surprising electrical properties.

The graphite crystal is another example of a valence crystal. Its structure and properties are discussed in Chapter 5.

4-2 Equivalent orbital model of polymers

The polythene molecule consists of CH_2 groups in a zig-zag configuration. The C atoms lie in a plane and each has its CH bonds in a plane normal to this plane and bisecting the CCC angle. It is a saturated molecule, a very long normal paraffin. In chemical terms the CC bonds connect the C atoms and with the CH bonds give approximate tetrahedral bonding around the C atoms. For a discussion of the possible symmetry groups of polymers see [1].

The equivalent orbital wavefunction [2] is based on this description. It assigns pairs of electrons to the localized inner shell (1s) of each C, to the CC bonds and to the CH bonds. In the ground state all these orbitals are doubly-occupied. The matrix elements of the Fock operator, H_{eff}, with respect to these orbitals are governed by the permutation symmetry of the orbitals. The unit cell contains two C atoms and four CH bonds but, by using the glide plane which brings one C atom into the next as the basic symmetry operation instead of a translation, we can work with the smaller repeating unit which has one C atom and two CH bonds.

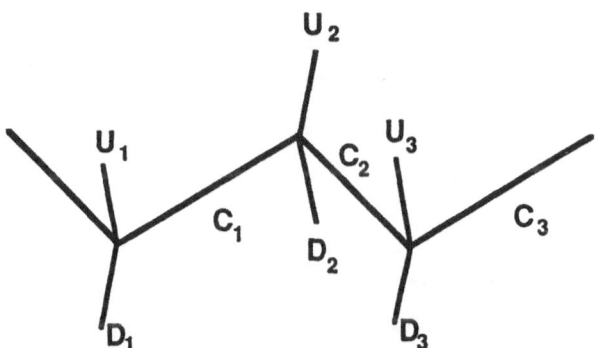

Fig.1 Localized equivalent orbitals for polythene

Figure 1 shows the relation of these orbitals to their neighbours. The orbitals are represented by lines. It is convenient to label the nth CC orbital as C_n with its up and down CH orbitals U_n and D_n. The diagonal matrix elements are

$$\alpha = \int C_n H_{eff} C_n d\tau, \quad \rho = \int U_n H_{eff} U_n d\tau$$

and the nearest neighbour elements (as determined by at least one atom in one bond being a nearest neighbour of one in the other) are

$$\beta = \int C_n H_{eff} C_{n+1} d\tau, \quad \sigma = \int U_n H_{eff} D_n d\tau$$

$$\gamma = \int C_n H_{eff} C_{n+2} d\tau, \quad \psi = \int U_n H_{eff} C_n d\tau$$

$$\lambda = \int U_n H_{eff} D_{n+1} d\tau, \quad \mu = \int U_n H_{eff} U_{n+1} d\tau$$

$v = \int U_n H_{eff} C_{n+1} d\tau$

All the more distant elements are assumed to be zero because of the localization of the orbitals.

The matrix of H_{eff} can now be described, in partitioned form, as consisting of rows that are similar to one another:

..O C^T B^T A B C O..

.. C^T B^T A B C ..

where T is the transpose operator, O is the zero matrix and the other submatrices are

$$A = \begin{pmatrix} \alpha\psi\psi \\ \psi\rho\sigma \\ \psi\sigma\rho \end{pmatrix}$$

$$B = \begin{pmatrix} \beta\psi\psi \\ \nu\mu\lambda \\ \nu\lambda\mu \end{pmatrix}$$

$$C = \begin{pmatrix} \gamma\nu\nu \\ 000 \\ 000 \end{pmatrix}$$

The occupied molecular orbitals, defined in terms of the localized orbitals, are the eigenvectors of this matrix. The use of the phase difference method of Appendix 2 is the first step in the process of diagonalization. It means introducing coefficients for all the orbitals in the nth cell with the same phase factor, $e^{in\theta}$. The eigenvalue equation now reduces to the diagonalization of:

$A + Be^{i\theta} + B^T e^{-i\theta} + Ce^{2i\theta} + C^T e^{-2i\theta} =$

$$\begin{pmatrix} \alpha+2\beta\cos\theta+2\gamma\cos 2\theta & \psi(1+e^{i\theta})+\nu(e^{-i\theta}+e^{2i\theta}) & \psi(1+e^{i\theta})+\nu(e^{-i\theta}+e^{2i\theta}) \\ \psi(1+e^{-i\theta})+\nu(e^{i\theta}+e^{-2i\theta}) & \rho+2\mu\cos\theta & \sigma+2\lambda\cos\theta \\ \psi(1+e^{-i\theta})+\nu(e^{i\theta}+e^{-2i\theta}) & \sigma+2\lambda\cos\theta & \rho+2\mu\cos\theta \end{pmatrix}$$

The structure of the last two rows and columns shows that one solution can be found immediately. This is the band constructed from the antisymmetric combinations of the two CH bonds on each atom and has the energy

$\varepsilon = \rho - \sigma + 2(\mu-\lambda)\cos\theta$

The other two bands have energies given by

$$\varepsilon = \alpha+\rho+\sigma+ 2(\beta+\mu+\lambda)\cos\theta+ 2\gamma\cos2\theta \pm \{[\alpha-\rho-\sigma+(\beta-\mu-\sigma)\cos\theta+2\gamma\cos2\theta]^2 +4\psi\nu(\cos\theta+\cos2\theta)\}^{1/2}$$

These three bands arise from the use of the glide plane to move one CH_2 group to the next. The strict translation period is twice as long. The effect of this is to double the number of variables in unit cell and so the number of bands is doubled. This is achieved, using the results above, by plotting the bands from 0 to $\pi/2$ and then folding each band back by using $\pi-\theta$ as the variable. These two bands will then "stick" at the point $\pi/2$. Sticking of bands is frequently the result when a glide plane or a screw axis is a basic symmetry operation in the structure.

To complete the discussion we need the numerical values of the integrals. In principle, these can be evaluated from approximate forms of the localized orbitals. In practice, it is more usual to estimate them by comparison with experiment. Typical values (Honegger *et al* [3]) are, in eV:

$\alpha = -17.50$ $\beta = -2.89$ $\gamma = 1.0$ $\rho = -16.95$ $\sigma = -2.89$
$\psi = -2.89$ $\lambda = 1.0$ $\mu = -0.5$ $\nu = -0.5$

With these values the shape of the bands is shown in Figure 2.

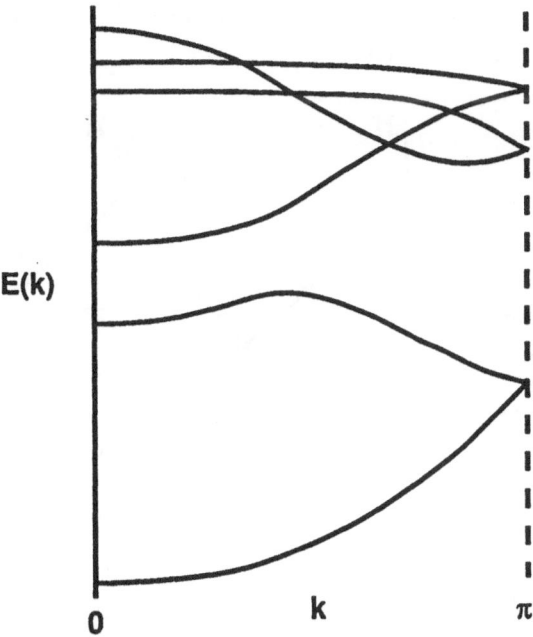

Fig. 2 Bands of polythene

From these bands, also, the density of levels as a function of energy can be calculated and this should be directly related to the observed density of ionization potentials in the photoelectron spectrum.

4-3 The model of trans-poly-acetylene

The trans-poly-acetylene (tpa) molecule has a very similar structure. Its C atoms also have a planar zigzag configuration. At each C there is now only one CH bond which is also in the molecular plane. The CC bonding is now unsaturated. The σ bond is localized but the π electrons are delocalized.

Before considering the complications of the actual structure of this molecule it is convenient to treat it in an idealized geometrical structure. This gives the CC bonds alternately long and short lengths. The loss of the glide plane symmetry means that the bands no longer stick. In each repeating unit there are now two C atoms each with its CH bond, its CC σ bond to its neighbour and a π orbital. The relations with its neighbours are shown in Figure 3.

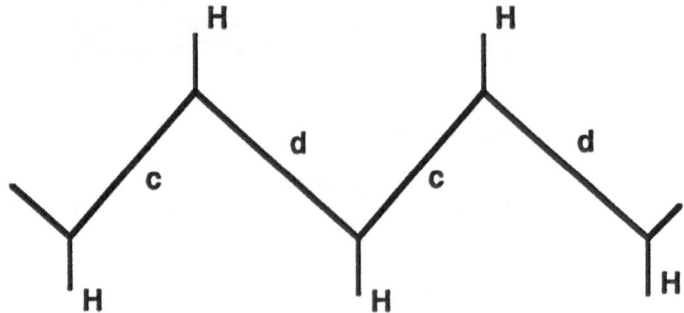

Fig. 3 The trans-poly-acetylene molecule

The matrix elements of the Fock Hamiltonian H_{eff} are defined as
c= energy of a short CC bond
d= energy of a long CC bond
h= energy of a CH bond

p= interaction of adjacent CH and CC bonds

q= interaction of trans CH bonds

r= interaction of gauche CH and CC bonds

x= interaction of adjacent CC bonds

y= interaction of trans short CC bonds

z= interaction of trans long CC bonds

With these, the σ molecular orbitals can be found as eigenvectors of the infinite Fock matrix. By using the phase difference method this reduces to the diagonalization of the finite matrix

$$\begin{pmatrix} h & p+re^{-if} & q(1+e^{-if}) & r+pe^{-if} \\ p+re^{if} & c+2y\cos f & p+re^{-if} & x(1+e^{-if}) \\ q(1+e^{if}) & p+re^{if} & h & p+re^{-if} \\ r+pe^{if} & x(1+e^{if}) & p+re^{if} & d+2z\cos f \end{pmatrix}$$

where f is the phase difference from one cell to the next.

The π orbitals are treated separately since they have antisymmetric symmetry with respect to the molecular plane and so do not mix with the σ orbitals. The matrix elements needed are

α = energy of π electron

β = interaction across a short bond

γ = interaction across a long bond

δ = interaction of second neighbours

Again, the phase difference method reduces the infinite Fock matrix to a finite matrix

$$\begin{pmatrix} \alpha+2\delta\cos f & \beta+\gamma e^{-if} \\ \beta+\gamma e^{if} & \alpha+2\delta\cos f \end{pmatrix}$$

which has the two bands of eigenvalues

$\varepsilon = \alpha+2\delta\cos f\pm\sqrt{(\beta^2+\gamma^2+2\beta\gamma\cos f)}$.

The values of all these integrals have been obtained [4] by fitting the energies to those obtained from an *ab initio* calculation of this molecule (Yamabe *et al* [5]). This gives, in eV:

c=-32.6, d=-17.4, h=-28.75, p=-4.5, q=-4.425, r=0.29, x=-4.2, y=-3.225, z=0.605, α=-5.675, β=-5.625, γ=-1.425, δ=-0.738.

The results show a good fit to all the calculated bands. The shape of the bands is given in Figure 4. The heavy lines indicate the pi bands.

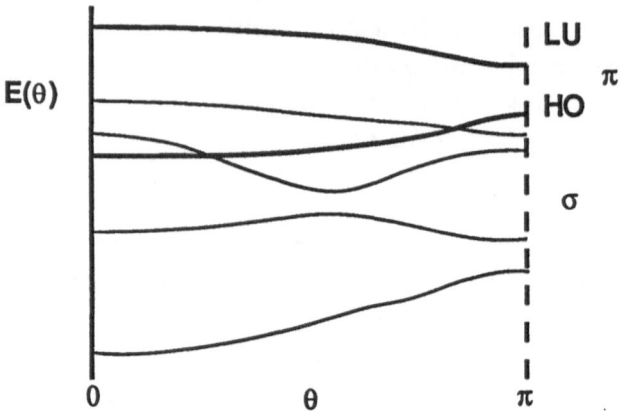

Fig. 4 Band structure of trans-poly-acetylene

Experimental information about this molecule is limited but it indicates that the band gap is about 1.4eV. The value above is 7.45eV and clearly needs some adjustment. This model treats the molecule as infinite in length. The real molecule has a finite length and this will impose end effects on the calculation. These are expected to have little effect on the shapes of the bands. Further aspects of this molecule are discussed in 4.10 and 4.11 below.

4-4 The diamond crystal

Diamond has a trigonal lattice. The three basis vectors e_i are equal in length and at equal angles with each other. Its unit cell contains one C atom which is bonded tetrahedrally to four others at the apex O of the cell and at the ends of its three basis vectors. It is easier to visualize the structure as cubic since the central atoms form one fcc lattice, and those at the cell vertices form another interlacing fcc lattice. This subdivision also demonstrates that the lattice is alternant.

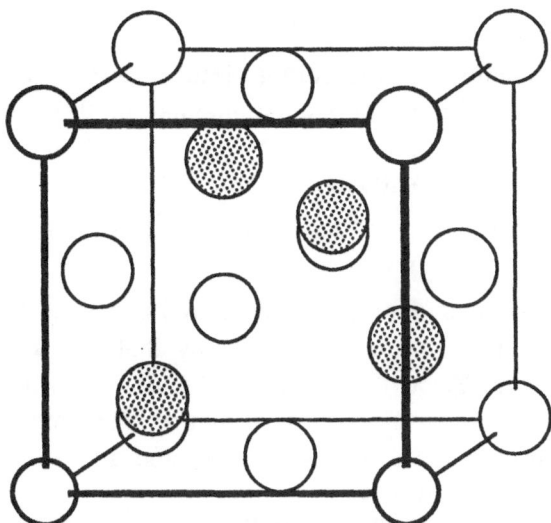

Fig. 5 Structure of diamond

Figure 5 shows the central atoms shaded. If these atoms were a different species this would be the zincblende lattice. The CC bonds can be described by localized equivalent orbitals [6] similar to those for the polythene molecule. These will be identical with one another except for their location and orientation. To set up the Fock matrix we require some matrix elements:

$$\alpha = \int \chi_0{}^* H_{eff} \chi_0 d\tau,$$

$$\beta = \int \chi_0{}^* H_{eff} \chi_n d\tau \qquad \text{n is a nearest neighbour of o}$$

$$\gamma = \int \chi_0{}^* H_{eff} \chi_t d\tau \qquad \text{t is a trans second neighbour of o}$$

$$\delta = \int \chi_0{}^* H_{eff} \chi_g d\tau \qquad \text{g is a gauche second neighbour of o.}$$

The occupied crystal orbitals (CO) are linear combinations of these localized equivalent orbitals and the symmetry of the lattice determines most of their coefficients. Each CO has a constant phase difference from one cell to the next along one of the base vectors. It is given a vector label containing these phase differences:

$$K = fE_1 + gE_2 + hE_3$$

where the reciprocal bases E_i satisfy

$$E_i \cdot e_j = \delta_{ij}$$

The CO satisfies the Bloch phase difference relation

$\psi(r+e_1) = e^{if}\psi(r)$

with similar relations in the other lattice directions. This is the generalization of the phase difference method to three dimensions. We can construct symmetry orbitals from this under the translation group. From the starting orbital $\chi_0(r)$ is obtained the orbital:

$\psi_0 = \chi_0(r) + e^{if}\chi_0(r-e_1) + e^{-if}\chi_0(r+e_1) + e^{ig}\chi_0(r-e_2) + e^{-ig}\chi_0(r+e_2) + e^{ih}\chi_0(r-e_3) + e^{-ih}\chi_0(r+e_3) + \ldots$

where the sum extends over all bonds parallel to the first one with the appropriate phase factor. It is easy to verify that this satisfies the Bloch relation given above. Since there are four bonds in unit cell, there will be four of these symmetry orbitals. The matrix elements of the effective Hamiltonian between these can now be evaluated and are:

$H_{00} = \alpha + 2\gamma\{\cos f + \cos g + \cos h\}$

$H_{11} = \alpha + 2\gamma\{\cos f + \cos(g-f) + \cos(h-f)\}$

$H_{22} = \alpha + 2\gamma\{\cos g + \cos(h-g) + \cos(f-g)\}$

$H_{33} = \alpha + 2\gamma\{\cos h + \cos(f-h) + \cos(g-h)\}$

$H_{01} = 2\beta \cos f/2 + 2\delta\{\cos(g-f/2) + \cos(h-f/2)\}$

$H_{02} = 2\beta \cos g/2 + 2\delta\{\cos(h-g/2) + \cos(f-g/2)\}$

$H_{03} = 2\beta \cos h/2 + 2\delta\{\cos(f-h/2) + \cos(g-h/2)\}$

$H_{12} = 2\beta \cos(g-f)/2 + 2\delta\{\cos(g+f)/2 + \cos(h-f/2-g/2)\}$

$H_{13} = 2\beta \cos(h-f)/2 + 2\delta\{\cos(h+f)/2 + \cos(g-h/2-f/2)\}$

$H_{23} = 2\beta \cos(h-g)/2 + 2\delta\{\cos(h+g)/2 + \cos(f-h/2-g/2)\}$

The four valence bands are obtained as the eigenvalues of this matrix. They will be functions of **K**. The eigenvalue equation:

$|H_{rs} - \varepsilon(K) \delta_{rs}| = 0$

can not be solved explicitly in general but can be in certain directions. Thus, in the direction (111) when f=g=h, the equation yields a doubly degenerate root

$\varepsilon = \alpha - 2\beta + 4\gamma - 2\delta + 2(\gamma-\delta) \cos f$

and the two roots

$\varepsilon = \alpha+2\beta+2\gamma+2\delta+(4\gamma+2\delta)\cos f \pm 2\{[\beta+\chi+\delta+(\delta-\gamma)\cos f]^2 + 3(\beta+2\delta)^2 \cos^2 f/2\}^{1/2}$

Similarly, along the direction (110) when h=0, f=g, the roots can again be found. They are the degenerate pair:

$\varepsilon = \alpha - 2\beta + 2\gamma + 4(\gamma-\delta) \cos f$

and the two

$\varepsilon = \alpha+2\beta+2\gamma+4(\gamma-\delta) \cos f \pm 4(\beta+2\delta) \cos f/2$

In particular these give, at the point (000), the triply degenerate

$$\varepsilon = \alpha - 2\beta + 6\gamma - 4\delta,$$

which is the uppermost level of all the bands, and the lowest one, which is

$$\varepsilon = \alpha + 6\beta + 6\gamma + 12\delta$$

The form of these bands is shown in Figure 6 for diamond.

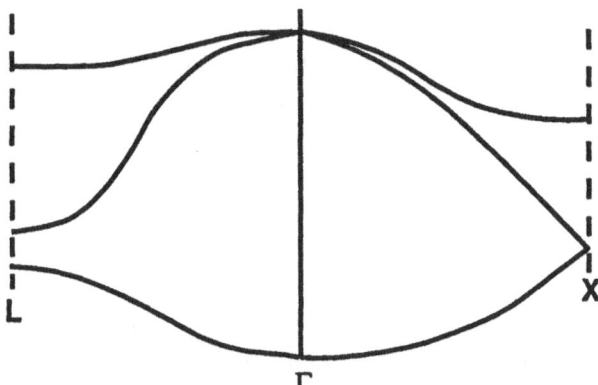

Fig. 6 Valence bands for diamond

The values taken for the integrals (in eV) are shown in Table 3. These values are obtained by comparison with the results of direct calculations of the band structure by other methods. It is found that a fair measure of agreement can be obtained when quite different methods are compared though there is no absolute agreement. A full discussion has been given by Levin [7].

Table 3. Integrals for the valence bands of diamond type solids

	α	β	γ	δ
C	-21.67	-1.86	0.607	-0.442
Si	-4.81	-1.36	0.34	-0.068
Ge		-1.46	0.3	-0.03

This technique gives the valence bands alone. To obtain the conduction bands the entire calculation has to be repeated using the anti-bonding orbitals between the C atoms. This exercise can be performed quickly by using the alternant property of the diamond lattice. Since the anti-bonding orbitals are anti-symmetrical under inversion at the mid-point of the bond, they have opposite phases on the two atoms and these phases need to be selected systematically. Now, if the four anti-bonding orbitals involving one C atom are given the same sign there, then all its nearest neighbours will have the opposite sign. The signs of every atom in one alternant set will be positive and those in the other will be negative so that the phases over the entire lattice will be specified. The matrix elements will then be exactly as before in form and so will the eigenvalues except that the values of the integrals will be different. These integrals can also be obtained by comparisons with other information.

4-5 Elastic constants for diamond

The reality of the valence nature of the bonding in the diamond crystal is emphasised by considering its distortions. In the study of molecular vibrations the role of bond-stretching and bond-bending forces have been found to be dominant. Often, these force constants for one kind of bond will transfer to other molecules containing the same bond with a high degree of accuracy. This representation of the elastic forces within a valence crystal is different from the representation derived from the Born theory which uses as the basis the cartesian displacements of individual atoms from their equilibrium positions. Any constancy of the force constants is hidden in such a treatment.

An application of this valence force field to the diamond crystal has been given by Musgrave and Pople [8]. In addition to the constants for stretching, k_r, and bending, k_θ', they introduce interaction constants k_{rr} for the effect of one stretch on the stretch of its nearest neighbour, $k_{r\theta}$ which couples the stretching and bending of a bond and k_θ'' which couples different angle-bendings around an atom. With these 5 constants they have included all the effects of its first and second neighbours on an atom.

From these parameters the potential energy can be evaluated and the entire vibrational spectra derived. The calculation is similar to the electronic structure calculation above though there are more bands to be included. By consideration of the selection rules, it becomes clear that the Raman spectrum will contain only one frequency. This is given by:

$$\nu = \frac{1}{2\pi c}\sqrt{\frac{8\alpha}{m}}, \quad \alpha = (k_r - 2k_{rr} - 8k_{r\theta} + 8k_\theta").$$

Another use for this potential energy is to derive the macroscopic elastic constants. These have the values:

$c_{11} + 2c_{12} = 1/2a(k_r + 6k_{rr}) = 3K$

$c_{11} - c_{12} = 3/a(2k_\theta' - k_\theta")$

$$c_{44} = \frac{3[(k_r - 2k_{rr})k_\theta" - 2k_{r\theta}^2]}{a[k_r - 2k_{rr} + 8(k_\theta" - k_{r\theta})]}$$

where a is the CC distance. K is the bulk modulus and c_{11}, c_{12} and c_{44} are the three independent elastic constants for a cubic lattice.

Unfortunately the various experimental measurements of the elastic constants of diamond are not in agreement so that this theory cannot yet be verified directly.

4-6 A vacancy in diamond

One of the effects of high energy radiation on diamond is to create a new absorption band. This has been ascribed to the creation of a vacant site by removal of a C atom. To investigate this possibility, Coulson and Kearsley [9] have considered a model of this vacancy.

They start from the valence model where each C atom has four directed valence orbitals which usually overlap with those of their nearest neighbours to constitute the bonding equivalent orbitals. The vacant site implies that four neighbours will each have one such directed orbital with one unpaired electron. These four orbitals will combine to give four molecular orbitals. One will be fully symmetric A_1 and the others span a triple-degenerate state T_2. There are four electrons (or possibly three or five, depending on external circumstances) to be allocated to these. They show that the results always involve degenerate states so that some Jahn-Teller distortion is to be expected. They give an approximate calculation of the expected excited states and show some relation to the experimental absorption band.

Very similar considerations are required when the electronic structure of the diamond surface is examined. The geometry of the surface layer will distort to enable the "dangling" bonds to link with one another.

4-7 Semi-conductors

The germanium crystal is a good example of a semi-conducting solid. This property is closely related to the fact that the gap between its full valence band and its empty conduction band is only 0.67 eV. The significance of the crystal orbitals is that they are the description of the electrons in the crystal which most nearly represents them as independent of one another. This encourages us to apply to them some results that have been derived for independent electrons in a band. One of these is the use of the Fermi-Dirac distribution to determine the effect on the occupation of the orbitals of a raising of the temperature.

The Fermi-Dirac distribution is written as

$$f(E) = \frac{1}{1 + e^{(E-F)/kT}}$$

where $f(E)$ is the probability that a one-electron state of energy E will be occupied by an electron. The parameter F in this is the Fermi energy; it is a chemical potential for the electrons. When this is applied to the region of the gap with T>0 it shows how the occupied band, at its highest point, is no longer fully occupied and the conduction band is not now completely empty. F is determined by the condition that the total number of electrons is constant. For very small T its value is exactly at the mid point of the gap.

The effect of temperature (T>0) is then to put some electrons into the bottom of the conduction band, E_e, where they can carry negative charge throughout the crystal. The empty levels at the top of the valence band, E_h, can also carry charge by moving the hole around. This hole behaves as a positive charge carrier. Thus the material has acquired carriers and will be conducting.

To turn the distribution into a density of charge the probability has to be multiplied by the density of states. For small temperatures this can be approximated by one constant factor for each band and the distribution function can be simplified so that we have, for the number of conduction electrons:

$n(T) = N_e e^{(F-E_e)/kT}$

and, for the holes:

$p(T) = N_h e^{(E_h-F)/kT}$

where N_e and N_h are the constant densities. We note that the product is independent of F

$pn = N_h N_e e^{-G/kT}$.

where $G=E_e-E_h$ is the energy gap. Semi-conductors of this type are called intrinsic semi-conductors.

4-8 Impurities and doping

The major step in the exploitation of semi-conductors came when it was realized that intrinsic semi-conductors could be modified by adding other atoms so that their properties changed significantly. An atom, added to germanium or silicon, which has more than four valence electrons will have extra electrons not required by the bonding to its neighbours. Similarly, if the valence is less than four, it will have insufficient electrons to form these bonds. Thus if a P atom is added to a silicon crystal its extra electron may be loosely attached to its nucleus at 0°K (see the H-atom model in 2-11) but, with some thermal disturbance at higher temperatures, it will donate its extra electron to the crystal and this will be accomodated in the conduction band. The crystal will then be conducting even at quite low temperatures. If the distribution of P is uniform and the distribution of extra electrons is also uniform there will be no net charge anywhere. This is n-type material since its carriers are negative. Similarly the addition of boron atoms, which will accept electrons and create holes in the valence bands, will produce p-type material with positive carriers.

The effect of this doping of the Si is to change the Fermi level. It is no longer mid way between the conduction and valence bands but is now raised in n-type and lowered in p-type material so that, near 0°K, the distribution reproduces the correct number of carriers. The consequence of this is that, at higher temperatures, the number of the main carriers is changed very little but some carriers of the opposite sign are also created. Their numbers are given by the approximate relations, for n-type,

$$n = D, p = N_e N_h e^{-G/kT}/D,$$

where D is the concentration of donors and G is the gap energy. Similarly, for p-type, the numbers are:

$$p = A, n = N_e N_h e^{-G/kT}/A$$

where A is the concentration of acceptors. These are called extrinsic semi-conductors.

4-9 The np junction

It is interesting to consider the behaviour of material which is n-type on one side of an interface and p-type on the other side. Since the n-type has an excess of electrons and p-type a deficiency there is a flow of electrons from one to the other. This creates layers in each material near the interface where there is now a net space charge. This produces a double layer whose potential acts to halt the flow.

This qualitative description will now be made more precise. We assume that the depleted layer in the n-type has width x_n and density eD (e is the proton charge) while the corresponding layer in the p-type has width x_p and density -eA. These charge clouds will change the potential energy of the system. By integrating the Poisson equation

$$\frac{d^2\phi}{dx^2} = 4\pi\rho$$

through the two layers it is easily seen that the potential has two parabolic segments which together form a step from one potential level to another. This energy change modifies the heights of both conduction and valence bands through the junction region as in Figure 7.

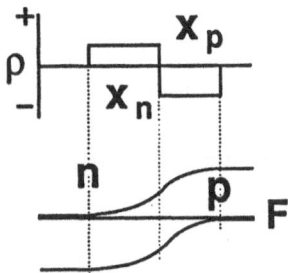

Fig. 7 Charge density ρ and bands in np junction

The Fermi level must be a constant for the entire material so it can be seen that the effect of the potential step is to give, far from the junction, the correct n or p type behaviour. It is as if the zero of energy in the two materials differs by a constant potential difference produced by the double layer. The situation closely resembles that in the electrolytic cell where a double layer forms and creates a back potential.

The interesting aspect of this junction is its reaction to an imposed electric field. If the field is in the forward direction then the Fermi energy is raised in the n-type and lowered in the p-type. Electrons then flow easily into the p-type and holes into the n-type. The carriers are increased and the current is enhanced. On the other hand, when the field is applied in the opposite direction the carriers experience barriers which have been heightened so that the flow is strongly diminished. This is the rectification property of such junctions.

4-10 The pnp transistor

The next stage in the technology of semi-conductors is the preparation of material containing two junctions. The pnp material forms the prototype of a transistor. Its behaviour apes that of an electronic valve and has now replaced it. It has three electrodes, emitter, base and collector respectively, and the potential differences between these are controlled externally. Between the emitter and the base there is a pn junction so the action is exactly as above. A large potential difference in the forward direction will produce a strong current. These carriers meet the reverse (np) junction and, if this junction is biassed in the reverse direction relative to the base, the current will again be multiplied. The result is an amplification of the current which is very sensitive to the potential difference between the base and the emitter. Current ratios of up to 60 are found.

Material like this need not be prepared in isolation and then assembled. It can be prepared from pure Si films by adding the correct doping locally. This film method can be done on a very small scale on a chip so that elaborate effects can be obtained efficiently and cheaply.

4-11 Excited states of trans-poly-acetylene

The polyenes tend to have trans configurations. Their long chain limit, poly-acetylene, has trans and cis forms though the trans is favoured thermodynamically. The infinite polymer, trans-poly-acetylene, has been studied for some time. It was the object of a paper by Lennard-Jones [10] where he introduced the ideas of compression energy and a resonance integral β depending on distance into the simple Hückel theory of hydrocarbons. He gave expressions for the total pi energy when the bond lengths alternate between long and short. A modern elaboration of this model has been given above in section 4.4.

The simple Hückel theory of the polymer with equal CC bond lengths obtains a band structure with the occupied orbitals extending up to $\varepsilon = \alpha$ where the unoccupied orbitals begin. There is, then, degeneracy between the highest occupied and lowest unoccupied molecular orbitals (homo and lumo). By the Jahn-Teller theorem the system will distort to resolve this. The alternation of bond lengths serves this purpose since it creates a gap between the homo and lumo. In this context the observation that a linear metal is not possible is due to Peierls [11] and this distortion is sometimes called a Peierls distortion. The conclusion has to be that the variation of β with distance is essential to a correct description. The simplest possible variation is the linear one:

$\beta = \beta_0 - \beta_1(R-R_0)$

where R_0 is a standard CC distance. A better representation is the exponential:

$\beta = B\, e^{-R/A}$

where A is a distance. The change of distance also changes the σ bond energy so a term in the total energy must reflect this. A simple form is:

$\Delta E_\sigma = 1/2\, K\, (R-R_0)^2$

This is exactly the form used by Lennard-Jones with R_0 the single CC length and K its force constant.

Pople and Walmsley [12] discussed one kind of excited state possessed by the polymer. In this, one electron is raised from the valence band to a localized level on one atom with $\varepsilon = \alpha$. On one side of this atom the double bonds alternate on the trans zigzag while on the other they have reversed. The state is a "wall" between two stretches of regular alternation. It is very similar to the Bloch wall, between two domains with different orientations of magnetic polarization, in ferromagnets. Evidence was soon found from esr spectra to confirm the existence of such states.

4-12 The SSH model of a soliton

The theory of the infinite polymer received a considerable boost from the experimental preparation of sheets of the material by Shirakawa *et al* [13]. In order to understand its properties and to predict new ones Su, Schrieffer and Heeger [14] (SSH) developed a model for the pi electron behaviour which has been of great value. It builds on the results above though it translates them into a different symbolism.

In the SSH theory the second quantization operators (see Appendix 4) are used. Thus $c_r^{\dagger}$ creates a π electron on the rth atom and c_r annihilates one there. The Hückel-like Hamiltonian can then be rewritten as:

$$H_h = \sum_r \left(\beta(R_r)c_{r+1}^{\dagger} \; c_r + \beta(R_r)c_r^{\dagger} \; c_{r+1} \right)$$

They use the linear approximation for the variation in β so that:

$\beta(R_r) = \beta_0 + \beta_1(u_{r+1} - u_r)\sqrt{3}$

where u_r is the displacement, in the direction of the molecular axis, of the atom r from its averaged position with all bonds of equal length. The full Hamiltonian for the system is then:

$$H = H_h + K/2 \sum_r (u_{r+1}-u_r)^2 + M/2 \sum_r \acute{u}_r^2$$

where M is the mass of the CH_2 group and $\acute{u}_r$ is the velocity of the atom r. This Hamiltonian with its related equations constitute the SSH model. It advances beyond Hückel theory in allowing dynamically for the variation in integral values with CC length and the motion of the atoms.

The infinite molecule has two ground states of equal energy. In one all the double bonds are on the left of the upward vertices and in the other they are all on the right. Since the system is periodic all the displacements have the same magnitude but alternate in sign so that:

$u_r = (-1)^r u_0$

Thus the two states can be described by

$u_r = \pm (-1)^r u_0$.

SSH consider that the Pople-Walmsley model of a lone electron forming a wall between two regions of different alternation is too localized. They have used an extended wall containing several atoms. The trial displacements which they use have the form

$\psi_r = u_0 \tanh(r/a)$.

By inserting plausible values for the various constants and minimizing the energy they find that a=7 and the excitation energy is 0.4eV. This excitation is free to move over the length of the polymer without change of shape. It is a neutral soliton but it has a spin moment. By adding a second electron with opposite spin a charged soliton is formed which has a zero spin and can be a carrier.

The significance of these solitons is that when the molecule is doped by appropriate substituents these excited states are created. The charged solitons are thought to be the carriers of the current in these semi-conductors.

References

[1] Hall G G 1967 Applied Group Theory, Longmans London
[2] Hall G G 1951 Proc Roy Soc A205: 541
[3] Honegger E, Yang Z and Heilbronner E 1984 Croatia Chem Acta 57: 967; Heilbronner E 1977 Helv Chim Acta 60: 2248
[4] Hall G G 1987 Synthetic Metals 17: 123
[5] Yamabe T, Tanaka K, Teramae H, Fukui K, Imamura A, Shirakawa H, and Ikeda S 1979 J Phys C 12: L257
[6] Hall G.G. 1952 Phil Mag 43: 338; 1958 Phil Mag 29: 429
[7] Levin A A 1977 Solid State Quantum Chemistry, McGraw-Hill New York
[8] Musgrave M J P and Pople J A 1962 Proc Roy Soc A268: 474
[9] Coulson C A and Kearsley M J 1957 Proc Roy Soc A241: 433
[10] Lennard-Jones J E 1937 Proc Roy Soc A158: 280
[11] Peierls R E 1955 Quantum Theory of Solids, Oxford
[12] Pople J A and Walmsley S H 1962 Mol Phys 5: 15
[13] Ito T, Shirakawa H and Ikeda S 1974 J Polymer Sci: Poly Chem Ed 13: 11
[14] Su W P, Schrieffer J R and Heeger A J 1979 Phys Rev Lett 42: 1698

Chapter 5 Metals

5-1 Introduction

In the past, the interest of physicists in metals has focused on the properties of their relatively free electrons, including electrical and thermal conductivity, and their magnetic properties. More recently, it has appeared that the properties of small clusters of metal atoms could be even more interesting to chemists since they can be powerful catalysts.

In this chapter we begin with comments on the geometrical and electronic structures of small clusters and extend the discussion to show how these become large clusters. Finally, we look briefly at alloys.

5-2 Small clusters and the Jahn-Teller theorem

We consider small clusters of alkali metal atoms such as Li. When two of these atoms come together, the orbitals of their outer electrons, which were degenerate at large distances, become split. One, the bonding orbital, is the symmetrical combination of the valence orbitals and has the two electrons to fill it. The antibonding orbital is antisymmetric and is now empty. The binding energy, in a simple picture of the process of forming a bond, is due to the lowering of the energy of the bonding orbital. The transition from the long-range pair of singly-occupied orbitals to the short range doubly-occupied orbital is an example of the Mott transition in solid state theory.

When three atoms come together it might be expected that they would also try to be symmetrical. The equilateral triangle is the most symmetrical configuration available. But when the orbitals interact there will be a symmetrical orbital which is lowered in energy and a doubly-degenerate level which has higher energy. In the ground state the lower orbital will be doubly occupied and this degenerate orbital will have a single electron so the resulting state will be degenerate. By the Jahn-Teller theorem, a degenerate state will try to distort to remove the degeneracy. Here, the symmetry of the equilateral triangle will be lost and the system will prefer to be linear.

A similar fate befalls the Li_4 cluster. Its tetrahedral form is not stable nor is its square form because of Jahn-Teller distortions. The best form is that of a rhombus with 60° angle. With five Li atoms the pentagon cluster is unstable and so is the square with one atom above it. *Ab initio* calculations (Beckmann *et al* [1]) confirm these qualitative arguments and give magnitudes to the atomization energies of these clusters.

It is clear that the form of these clusters cannot be understood in terms of the classical idea of Li as a univalent atom. Furthermore, an analysis of the electron density of Li_2 in its equilibrium separation (Gatti *et al* [2]) shows that it has a maximum at the mid-point between the nuclei! (Non-nuclear maxima are most unusual in molecules.) The theoretical implication of this is that the atoms are not directly bonded but that there is a pseudo-atom at the mid-point, with a valence of two and a small negative charge, which binds the Li atoms. Similarly, in Li_4, the electron density shows that it has two pseudo-atoms, at the centres of the triangles, which are bonded together and to the three nearest Li atoms. Clusters of Na atoms (Cao *et al* [3]) exhibit the same forms and require similar pseudo-atoms to explain their electron densities.

5-3 Clusters of metal atoms

In recent years it has become experimentally possible to introduce metal vapour into a supersonic nozzle with an inert carrier gas and form metal clusters when the jet expands into a vacuum. The mechanism of cluster formation is not fully known but the indications are that the abundance of a cluster of a given size is related to the stability of that cluster. This technique has given a wealth of information about clusters.

The success of simple one-particle models of the nucleus in explaining the stability of certain nuclei has prompted the development of similar models of these clusters. One of these models is due to Knight *et al* [4]. It assumes that the valence electrons are in delocalized orbitals belonging to the cluster and that the potential in which they move can be represented by a simple spherical well. The Wood-Saxon potential (see Flügge [5]) is used. This has the form

$$V(r) = - \frac{U}{e^{(r-r_0)/a} + 1}$$

where U is the depth of the well, r_0 is its radius and a is the thickness of the surface layer in which the potential drops from outside to inside the sphere. The sequence of energy levels in this potential is different from those of the H atom or the spherical oscillator so the filling of electron shells occurs at different numbers of electrons. To relate to a specific metal the depth U is taken as its bulk binding energy and the radius is

$$r_0 = r_s N^{1/3}$$

where N is the number of valence electrons and r_s is the covalent radius of the metal atom. For Na the values are

U = 5.93 eV, r_s = 3.93 Bohr, a = 1.5 Bohr.

The eigenvalues for this potential can be easily calculated numerically.

The important feature of this potential for our purpose is the sequence of energy levels and their degeneracy. The theoretical order is 1s, 1p, 1d, 2s, 1f, 2p, 1g, 2d, 3s... By assigning electrons to these in pairs, to allow for spin, it is seen that the most stable clusters will be expected to be those with the following number of alkali atoms

$(1s)^2$: 2
$(1s)^2(1p)^6$: 8
$(1s)^2(1p)^6(1d)^{10}$: 18
$(1s)^2(1p)^6(1d)^{10}(2s)^2$: 20

and, similarly, we get after these 34, 40, 58, 68 and 70. These are exactly the "magic numbers" of Na atoms observed in the experimental results as the clusters with the largest populations in the jet.

The results can be made more quantitative. By adding the contributions from the various shells the total energy E(N) can be evaluated. The binding energy of the final atom is then

$\Delta(N) = E(N) - E(N-1)$

The change in this energy is

$\Delta_2(N) = E(N+1) + E(N-1) - 2E(N)$

and this becomes large when (N+1) begins a new shell. See Figure 1, which has been adapted from [4].

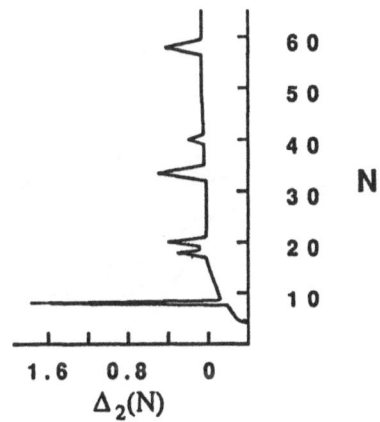

Fig. 1 Populations of Na clusters in a jet

This energy should be related to the observed population of clusters of size N in the experiments, according to certain theories. It does seem to correlate well with experiment. For a review of alternative theories and of experimental techniques and results see de Heer *et al* [6].

5-4 Larger clusters

As the size of the cluster increases so the constraints due to the Jahn-Teller theorem become less important. There are more geometrical degrees of freedom with which to make adjustments to split any degeneracy. These constraints are also greatly reduced when the atoms have two valence electrons. For reviews see Takasu and Bradshaw [7] and Burch [8]. The cohesive forces in these larger metal clusters are represented by Morse potentials between every pair of atoms. This potential falls off rapidly and leads to a form of close packing as the favoured structure. For the cluster of 13 atoms two forms are possible. Both have twelve atoms around a central atom. These are the icosahedron form and the cuboctahedron form, the first having a slightly lower energy. This size of cluster, which completes such a stable structure, has a low energy per atom. This is the first of the "magic" numbers due to close packing.

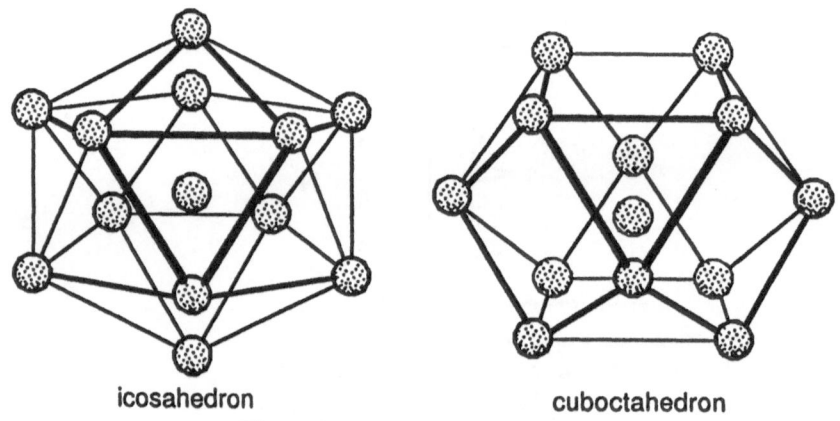

icosahedron   cuboctahedron
Fig.2 Clusters with 13 atoms

Outside these structures another layer can be added. This leads to the next magic number which is 55 and is another icosahedron. Eventually the structure must begin to form the bulk material. This is exactly what the cuboctahedron does, becoming a face-centred cubic, but the icosahedron has a five-fold axis and so cannot form any regular solid. A possible development of this is discussed in 6-10.

The importance of these clusters for catalysis is due to their surface to volume ratio which is large compared with that for planar surfaces. The details of their performance depend on the nature of the catalytic effect. In some reactions the area is the significant variable, in some it involves the geometry of a particular surface so that the area of surface with that geometry is important but in others it is the number of specific sites on the surface, such as edge sites, which is important. The behaviour of a cluster as a function of temperature is often of interest. Since the atoms have less cohesion than the solid the melting point of the cluster is lower, even as low as 1/2 T_m. It can happen that the outer part of a cluster is liquid while its inner core remains solid.

A different kind of cluster is formed when small molecules form a cluster around an ion. The strong electric field of the ion interacts with the dipole moments (or induced dipole moments) of the molecules to produce a stable nucleus for a large cluster. Small clusters are often thermodynamically unstable because most of their atoms lie on the surface and so are lightly bound. They are readily lost by evaporation. The presence of an ion enables the cluster to overcome this initial barrier to the formation of large clusters. The mechanism is important in the seeding of rain clouds.

5-5 Fractal clusters

Many growths show a common aggregation mechanism i.e. that particles 'stick' when they collide with the cluster at random points and so form larger clusters. This growth is not systematic nor orderly and results in open structures with many vacancies. One device for investigating these clusters is to relate N, the number of particles in the cluster, to its "radius of gyration", $R = \sqrt{<(r-<r>)^2>}$, where r is the vector position of a particle and the angular brackets imply taking mean values over the cluster.

When these variables, N and R, are plotted on log-log paper it often happens that there is a linear region. This occurs in the middle region, where N is not small and yet the density is not too high. The slope of this line is the fractal dimension of the clustering. It results in the relationship

$N = kR^D$

where D is the fractal dimension and k is a constant. In a true fractal situation the relationship would become more accurate as the scale became smaller. Here the fitting of a new atom into the cluster determines a minimum scale. The fractal character arises because of the open nature of the cluster.

With a flexible linear polymer, e.g. polythene in its liquid phase, in which the bond between monomers permits free rotation, the structure is fractal-like. For a rotation which preserves the tetrahedral angle the dimension is D=1.33 [9]. The linear system is behaving a little like a two dimensional system! Another well-known example is soot. This aggregates in a very open fashion and its dimension is D=1.55 so it is behaving more like a surface than a solid. In particular the ratio of surface area to volume will be much larger than for most materials. This property of large surface area can be very significant for certain catalysts and draws attention to their mode of preparation.

5-6 Corrosion

The process of corrosion is of such industrial importance that it cannot be ignored here although there is no space to develop the subject. A recent account has been given by West [10]. Corrosion happens to most metals when exposed to gases or liquids with which they can react. Even in dry air, if the temperature is high enough, many metals will oxidise. The more usual process of oxidation, however, is in the presence of water. A film of water on the surface of a metal contains some dissolved oxygen and will begin to dissolve the metal.

When water is in contact with a metal and an oxidant, such as air, an electrolytic cell is formed and electrons can pass through the system. In the solution the metal produces cations and atmospheric oxygen produces O anions. The ingredients for an oxide layer are present. As the water evaporates this layer will be deposited. The process of dissolving the metal is terminated by the formation of a double layer on the surface of the metal. This layer of ions in the water generates an electric potential barrier which prevents the further dissolving of the metal. If the ionic balance of the solution is disturbed or some other potential difference acts then the dissolving may proceed rapidly.

5-7 Metal cohesion

Solid metals tend to crystallize in one of the cubic forms. The alkali metals, for example, are all body-centred cubic. From the chemical point of view the natural explanation of their bonding is through the analogy with the π electron bonding of conjugated molecules. These metals have no underlying σ bond structure so the internuclear distances are larger than in their diatomics and the bonding is weaker. The bond order between the nearest neighbours in the alkali metals has been calculated as 0.23719 and for the next nearest neighbours 0.10811 (Löwdin [11]). The more distant neighbours have bond orders which become smaller in magnitude and alternate in sign so that these two contributions to the energy of binding will predominate.

The Morse potential is often used to represent realistically the potential between every pair of atoms. This has the form:

$$V(r) = D\{e^{-2\alpha(r-r_0)} - 2e^{-\alpha(r-r_0)}\}$$

where D is the dissociation energy reduced from the diatomic value because of the reduced bond order, r_0 is the equilibrium distance for the diatomic and α is a parameter governing the force constant. Since there are three constants in this expression it is not possible to scale it into an universal form. If, however, it is assumed that, for metals of the same type, there is a relation between the constants then it can be scaled for these metals. Thus using the single variable

$$x = \alpha(r-r_0)$$

and scaling the energy by D then, for any one crystal structure, the cohesive energy can be expressed in universal form. Similarly, the melting temperature scaled by D should be universal. From this, the relation, known as Trouton's law, follows that the cohesive energy is proportional to the melting temperature. This law is observed to hold for many metals.

For the close packed structures the energy per atom will be

$$E = 12D\{e^{-2x} - 2e^{-x}\} + 6D\{e^{-2x\sqrt{2}}k^2 - 2e^{-x\sqrt{2}}k\} + \ldots$$

where $k=e^{(1-\sqrt{2})\alpha r_0}$. Because of the second term, which depends on the value of k, this is not a universal form though it must be a close approximation to it since the first term must be the larger. The equilibrium distance between nearest neighbours will now depend slightly on k. For the bcc structures the second term is more significant so the scaling will not be so good an approximation.

5-8 Atom to metal transition

When metal atoms are sufficiently far apart their electrons are not free to migrate from one atom to the next and the typical delocalized behaviour of the metal electrons is absent. It is of interest to discover when this isolation breaks down and the usual metal properties appear. This matter has been investigated by Edwards and Sienko [12] and they give a criterion for this transition.

The Clausius-Mossotti formula for n, the high-frequency refractive index of a gas, is expressed as:

$$\frac{n^2-1}{n^2+2} = \frac{R}{V}$$

where R is the molecular refractivity, which is proportional to α, the molecular polarizability, and N_0 the Avogadro number (see Appendix 8)

$$R = \frac{4\pi N_0 \alpha}{3},$$

while V is the molar volume. This formula becomes meaningless when R=V since then n would be infinite. Thus the response of the gas to an electro-magnetic field in this model breaks down. Edwards and Sienko suggest that the result is the formation of the metal. Thus their criterion is that R>V will be metallic and R<V will not be metallic. The observations support this.

5-9 The Hume-Rothery rules for alloys

Two metals may be melted together and may or may not mix. Even if they are soluble it does not follow that they will form a homogeneous solid when they are cooled. If a homogeneous solid, an alloy, is formed it can be of various types. It is useful to distinguish several kinds. If a small amount of one is added to the other without disturbing its crystal structure this is a primary alloy. It is usually formed by substituting for one of the majority atoms by a minority one. If, on the other hand, the mixture adopts a new crystal structure it is a secondary alloy. Rules governing the stability and formations of these alloys have been formulated by Hume-Rothery [13].

Since the first essential is that the two metals should be soluble the first rule concerns this. It is:
Rule 1: When the metal atomic radii differ by more than 15% their solubility is small.
We have already seen from Pauling's rules that the radius ratio is an important variable for ionic crystals. It is not unexpected that it is needed here although its importance is usually explained in terms of the energy of distortion of the crystal rather than its geometry. Since several different crystal structures are found for metals this specific ratio is a broad generalization covering many different examples.

Metal atoms may differ considerably in electronegativity. If the difference is too large the atoms will form compounds rather than alloys. Thus the second rule points to the importance of the difference:
Rule 2: The greater the electronegativity difference of the metal atoms the less the tendency to form alloys
A closely related rule says that, if the two metal atoms have different valency, then the one with the higher valency will dissolve in the other but the lower valency atom will not dissolve in the higher.

The third rule determines the relation between the structure and composition of secondary alloys. It can be expressed in general form as:
Rule 3: The ratio of the total number of valence electrons to atoms is critical for certain structures.
The body-centred cubic lattice is a structure often found in metals and alloys. The alloy may be formed with the atoms placed randomly on the lattice points. The alloy, β-brass with a composition of approximately CuZn, is a typical example. If the number of atoms of the two metals is equal there will also be an ordered phase in which the atoms separate so

that all the atoms of one species are on one simple cubic lattice and all the other atoms on the other. This phase becomes the more stable one at lower temperatures and can be induced by annealing. The nature of the transition is discussed in Chapter 7. The ratio of valence electrons to atoms for this structure is 3/2. Other examples of ordered structures with this ratio are CuBe, AgCd, Cu_3Al and CoAl.

Another structure often found is hexagonal close-packed cubic. The prototype is $CuZn_3$ and others include Cu_3Si, Ag_3Sn and $FeZn_7$. These have the approximate electron to atom ratio 7/4. In some of these alloys the c axis is lengthened from its close-packed value. A third structure, found for γ-brass with the composition Cu_5Zn_8 , has a large, complicated unit cell containing 52 atoms. Its electron-atom ratio is close to 21/13. Other examples with the same ratio and the same structure include Cu_9Al_4, $Cu_{31}Sn_8$ and Co_5Zn_{21}.

These structures are not covalent compounds in the usual sense. The stochiometric ratios are only approximate and, as is obvious from the examples above, the bonding is not related to conventional valency. The explanation is to be found in the occupancy of the electronic energy bands. For these special electron-atom ratios a particular band of crystal orbitals becomes nearly fully occupied and this makes the structure especially stable.

References

[1] Beckmann H O, Koutecky J and Bonacic-Koutecky V 1980 J Chem Phys 73: 5182
[2] Gatti C, Fantucci P and Pacchioni 1987 Theoret Chim Acta 72: 433
[3] Cao W L, Gatti C, MacDougall P J and Bader R F W 1987 Chem Phys Lett 141: 380
[4] Knight W D, Clemenger K, de Heer W A, Saunders W A, Chou M Y and Cohen M L 1984 Phys Rev Lett 52: 2141
[5] Flügge S 1970 Practical Quantum Mechanics I, Springer Berlin
[6] de Heer W A, Knight W D, Chou M Y and Cohen M L 1987 Solid State Physics 40: 93
[7] Takasu Y and Bradshaw A M 1977 Chem Phys of Solids and Surf. 7: 59

[8] Burch R 1985 Catalysis 7: 149

[9] Kolb M, Botel R and Jullien R 1983 Phys Rev Lett 51: 1123

[10] Löwdin P O 1956 Phil Mag Supp 5: 1

[11] West J M. 1980 Basic corrosion and oxidation, Ellis Horwood Chichester

[12] Edwards P P and Sienko M J 1978 Phys Rev B17 2575

[13] Hume-Rothery W 1969 Structure of Metals and Alloys, Inst of Metals London

Chapter 6 Surfaces

6-1 Introduction

In earlier Chapters the energies of formation of some surfaces have been considered and certain aspects of surface structure and formation have been discussed. Since there are many important applications of our subject which depend critically on the details of the surfaces involved we now develop a more careful approach to surface properties. We begin by looking at the possible symmetries of a surface. The interesting possibility of surface structures which have local five-fold symmetry is also examined briefly.

Some chemical reactions are catalysed by bringing the reactants together on a suitable surface. As background to this catalytic action we look at the structure of the surface and its interaction with an approaching atom. When the surface is a metal, localized surface states of the metal may be involved in the chemisorption. Useful models for the discussion of this effect are discussed further in Appendix 5.

6-2 The surface lattice

Although there are 230 different space groups there are only 17 different two-dimensional lattice groups. This modest number makes it possible to outline their derivation one by one and so come to an understanding of the principles which govern all such enumerations of lattice groups.

The two-dimensional lattice is generated by two basis vectors. A first classification of the lattices is derived from the relations of these to one another. The possible relations of symmetry between the vectors **a** and **b**, their lengths a and b, and their contained angle θ can be described using the shape of the basic cell as:

I. Parallelogram a, b, θ

II. Rectangle a, b, $\pi/2$

III. Square b=a, $\pi/2$

IV. Hexagon b=a, $\pi/3$

Thus there are four systems of two-dimensional groups corresponding to the seven systems of three-dimensional crystals described in Chapter 1. For each of these the compatible point groups have to be considered. The system I has the least symmetry. If we take the origin as a typical lattice point the only point symmetry there, which is consistent with the lattice, is the inversion about that point. This means that there are exactly two surface groups in I depending on whether or not this

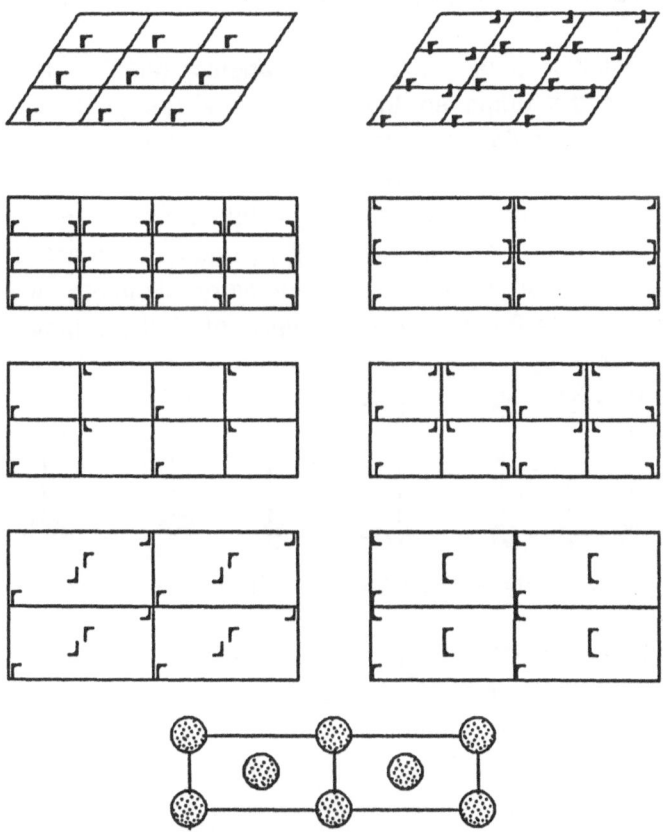

Fig.1 The surface groups with parallelogram or rectangle unit cells

inversion is present. As we have seen in 3-3, an inversion at each corner of the cell induces other inversions at the middle of the edges and the centre of the cell. Examples of these are shown at the top of Figure 1.

In the rectangular system II, reflections in planes through the two axes at each lattice point are possible as well as a centre of inversion. It is also possible to have a glide plane normal to the surface and intersecting it along one axis. If this axis does not pass through a base vector a face-centred lattice is produced. (In some classifications the set of face-centred lattices is taken as a separate system since its smallest unit cell is no longer a rectangle. Its basis vectors are one side and half the diagonal.) This produces three face-centred groups and a total of seven more groups. (We note that, if the arrangement within the rectangles is sufficiently unsymmetric and no plane or glide plane remains, the overall symmetry is reduced to that of one of the parallelogram groups.)

The square lattice differs from the rectangular one by allowing a four-fold rotation at the origin so that three more groups become possible which retain this rotation but may lose some of the other elements of symmetry of the square. One of these groups is a face-centred lattice. Similarly, the hexagonal lattice has a six-fold rotation around the origin and there are five more groups which retain at least a three-fold rotation but may lose some of the symmetry elements of the regular hexagon. The Figures 2 and 3 show examples of all these groups.

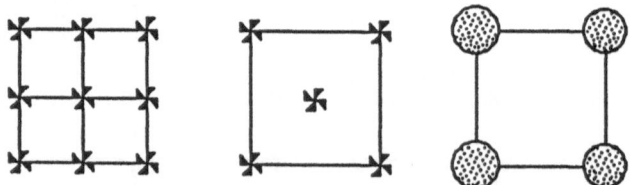

Fig. 2 The surface groups with square unit cells

Very similar arguments about the compatibility of various point groups, and their subgroups, with the lattice are used to show that, in three dimensions, the possible number of space groups is limited to 230 but, since this number is much larger, the arguments have to be applied in a more systematic way.

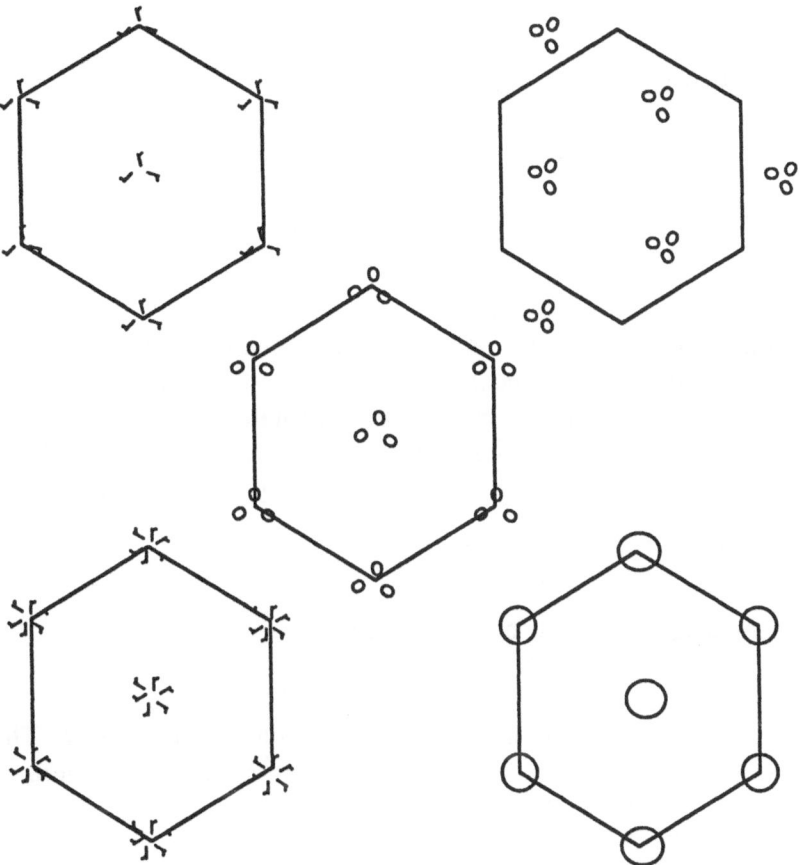

Fig. 3 The surface groups based on a hexagon

It is convenient also to have a notation to describe a layer of adsorbate on the surface if it forms a lattice which conforms to the surface lattice. This is done by describing the basic cell for the layer in terms of that for the underlying surface. Thus if the adsorbate has the same size and shape of lattice (possibly displaced in origin) then it is called (1×1). If it occupies a lattice whose unit cell has two basis vectors, one m times the length of the corresponding unit vector of the surface and the other n times the length of the other, then the notation will be $(m \times n)$. On a square lattice $(\sqrt{2} \times \sqrt{2})$ means that the layer is also square but its basis has been rotated so that the diagonals of the surface lattice are the sides of the new square. On a hexagonal surface $(\sqrt{3} \times \sqrt{3})$ means that the layer is hexagonal but rotated and enlarged so that each side of the hexagon is $\sqrt{3}$ times that of the surface hexagon. Examples of these surface structures and their identification using Low-Energy Electron Diffraction (LEED) have been given by Somorjai [1].

6-3 Energy of an atom approaching a surface

As an atom approaches the surface of an insulator it experiences a dispersion attraction from the nearest atoms. This should be summed up over the entire solid. The sum can be estimated for some simple situations by integration. Thus, for a monatomic solid with the density of atoms, ρ, and the LJ potential between the oncoming atom and an atom of the solid:

$$u(r) = -\frac{A}{r^6} + \frac{B}{r^{12}},$$

the energy as it approaches a planar surface will be:

$$U(z) = -\frac{\rho\pi A}{6z^3} + \frac{\rho\pi B}{45z^9}$$

where z is the length of the normal from the atom to the surface. This shows how the surface can attract atoms from outside to adhere to itself.

When an atom approaches a metal surface it should be expected, because of the greater freedom of the metal electrons, that a different formula will apply. The first attempt to estimate this effect was made by Lennard-Jones [2]. He assumed that the surface behaved as a macroscopic metal reflecting what was before it. Thus he added, to the Hamiltonian for the atom, extra terms for the image atom with reversed charges. From this, using perturbation theory, he deduced the formula:

$$U(z) = -\frac{e^2 <r^2>}{12z^3}.$$

Since $<r^2>$ can be related to the diamagnetic susceptibility of the atom he could predict the energies for inert gas atoms approaching some metals. His argument is now considered to be oversimplified since the metal cannot respond instantly to the motion of the electrons in the atom. An improvement to the formula was given by Bardeen [3] who inserted the factor

$$\frac{Ce^2}{2r_s\Delta + Ce^2}$$

where r_s is the radius of a sphere in the metal for one conduction electron, Δ is the ionization potential of the atom, e is the electron charge and C is a constant, C=2.6.

These formulae have been examined by Mavroyannis [4] and compared with some experimental results. A selection of his results is given in Table 1.

Table 1. Energy of an atom on a metal surface (cal/mole)

	z (Å)	Expt	LJ	Bardeen
Pt - He	2.70	265	890	230
Pt - Ne	2.98	330	2190	370
Pt - Ar	3.30	1320	5170	940
Pt - Kr	3.36	2110	7120	1240
Zn - Ar	3.24	1570	5465	1100
Cr - Ar	3.18	2090	5800	1240

This Table makes clear that the LJ relation overestimates the energy whereas the Bardeen relation, though much closer, underestimates it. Various improvements to these formulae have been suggested but no wholly satisfactory one has been found.

6-4 Clusters as surface models

The difficulties of performing accurate calculations on surface effects, where the crystal is semi-infinite, have led to the use of cluster models of the situation. In these models the crystal is represented by a finite cluster with the same geometry and composition as in the crystal. It would be logical to repeat the calculations with clusters of different sizes so that the effects of size can be estimated. These effects are known to depend markedly on the property of interest. In general, convergence to bulk properties would be expected only for clusters containing hundreds of atoms. Nevertheless some effects are not so sensitive and some can be extrapolated from much smaller clusters.

An early example of this was given by Suthers *et al* [5]. They investigated the relaxation of geometry at the surface of a LiH crystal using several small clusters. The distortion of the surface atoms from the crystal plane was clearly shown with the anions moving outward and the cations inward. Some disturbance was also found in the second and third layers.

6-5 Oxygen on graphite

Graphite is composed of layers each of which consists of conjugated hexagonal rings of C atoms in a two-dimensional honeycomb (Figure 4).

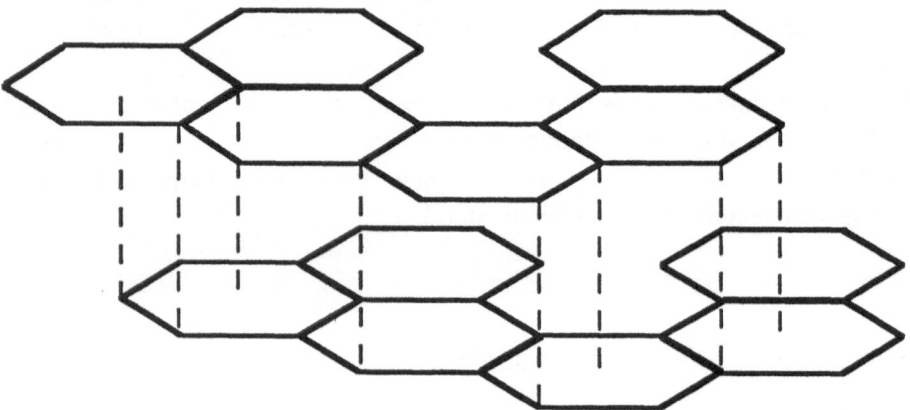

Fig. 4 The graphite lattice

The layer lattice is alternant and the two non-neighbouring sets in each layer are treated differently when these are stacked. From one layer to the next the hexagons are displaced so that alternate C atoms are directly above one another. The remaining C atoms are over the centres of the hexagons in the lower layer. Thus one alternant set has nearest neighbours in the layers above and below it which are in the same vertical line whereas, in the other set, each C has rings above and below it.

This layer structure is significant when the bonding is considered. The approximation of treating each layer as an infinite conjugated molecule is good but not the whole truth. The evidence from experiment (Shoenberg [6]) is that the Fermi surface (which divides occupied from empty orbitals) has the shape of a cigar indicating that there is some bonding from layer to layer. The introduction of a small Fock matrix element between those C atoms which are so close vertically will modify the electronic structure in the correct way (Johnston [7]) to give weak binding between layers.

Monolayers of molecular oxygen form on the surface of graphite in two different forms. There is a high density form and a low density form. In the high density form the molecules are lying normal to the surface in a hexagonal close-packed form. In the low density form the molecules are flat on the surface and pack together in a rectangular face-centred lattice.

The interactions between an oxygen molecule and the graphite are represented by a sum of dispersion and repulsion terms between each O atom and all the C atoms on the surface (Pan *et al* [8]). The result is a potential which reflects the periodicity of the graphite but with a rather small difference between its maximum and its minimum. Thus, when the monolayer is formed, it has a structure which is more determined by the non-bonding O-O interactions and is not entrained by the periodicity of the surface. On the other hand, when we consider the behaviour of small clusters of molecules on the surface, there is a tendency for the O-C potentials to be more important and individual molecules are pinned flat on the surface. These clusters, when sufficiently large, begin to form a two-dimensional monolayer of the low density form.

The result of the calculations is that a molecule which is normal to the surface has a minimum energy which is less attractive than the minimum energy of a molecule lying on the surface. Thus the transition to the high density form occurs when, with large clusters or multiple layers, minimizing the O-O interactions becomes more important. The calculations also give estimates of the temperatures when the transition between the two structures occurs and when the layer begins to melt.

6-6 Intercalation and graphite fluoride

The rather open ring system of graphite leads to one unusual property. Graphite can attract other substances inside itself. This is called intercalation. These intercalates form layers which lie between the graphite layers and may interact with them. This intercalation, in the example of fluorine, leads to the dissociation of the halogen molecule and the formation of saturated bonds between these atoms and the graphite.

There are several forms of graphite fluoride (for an extensive review see Watanabe *et al* [9]). In $(CF)_n$ every C atom has a F bonded to it. The aromatic character of the graphite is lost and its geometry reverts to that of a saturated hydrocarbon. In the top layer the set of C atoms which are not slightly bonded below are attacked by the F atoms while the intercalated fluorine then attacks the remaining C atoms from below. Thus the layer becomes puckered like a set of chair-form cyclo-hexane rings linked together. The surface symmetry is hexagonal. See Figure 5

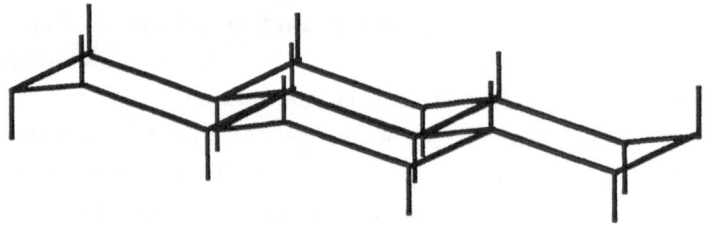

Fig. 5 The layer structure of $(CF)_n$

The second fluoride form $(C_2F)_n$ also starts by an attack on the set of surface C atoms which are not connected below. But, in this compound, the change to saturated bonding strengthens the bond to the next layer so that two layers of graphite remain attached and the remaining C atoms in the lower layer are attached to the other F atoms. The top layer has F atoms on its upward C atoms and the second layer down has the lower C atoms similarly bonded. The dissociation energy of the F_2 is the rate determining step in the fluorization process though there is also a barrier to the diffusion of the molecule inside the graphite or CF layers.

There is evidence also for other fluoride forms but their structures are less well established.

The industrial importance of this (CF) material is that it is used as the anode in the Li battery. The resulting battery is remarkably powerful for its size and very stable in voltage etc.

6-7 The McCreery-Wolken model of adsorption

Reactions on a surface differ from gas phase reactions in various respects. Since the reactants are in the same plane their probability of colliding is enhanced. Also, they may interact with the surface atoms so that the bonds that are to be lost may already be weakened or broken. Then there is the dynamic effect that the surface atoms can supply vibrational energy to facilitate a reaction or absorb it to stabilise the product. These various aspects have to be considered when a model of the attraction of a molecule to a surface is set up.

The McCreery and Wolken [10] model is for the interaction of hydrogen with a tungsten surface. This surface has full square symmetry. The model has to allow for the dissociation of the molecule, for the binding of each atom to the surface and for the movement of the atoms on the surface. They use the ideas of the LEPS treatment (see Appendix 4) which is so useful as a representation of the H_3 reaction energy surface. In effect four atoms must be considered, the two atoms of hydrogen and two surface atoms which could bind them.

The potential between an atom and the crystal surface cannot be described by the usual Morse function so they have defined a new function V(xyz). This has the form:

$$V = D(x,y)\left(e^{-2\alpha(z-z_0)} - 2e^{-\alpha(z-z_0)}\right)$$

where z is measured along the surface normal with z=0 at the surface, a is the side of the square and

$D(x,y) = D_0(1 + \delta\{\cos(2\pi x/a)+\cos(2\pi y/a) - A[\cos(2\pi x/a)-1][\cos(2\pi y/a)-1]\})$

$z_0 = z_m(1 + \varepsilon\{\cos(2\pi x/a)+\cos(2\pi y/a) - B[\cos(2\pi x/a)-1][\cos(2\pi y/a)-1]\})$

$\alpha = [0.02894/D(x,y)]^{1/2}$

In this, the parameters, D_0, δ, A, z_m, ε and B are constants whose values are fixed using some semi-empirical calculations. The square symmetry of this potential is obvious as is its representation of the corrugated shape of the surface.

With this potential it was possible to investigate the recombination of H atoms on the surface and the collisions of one atom with the surface where another H atom is adsorbed. The insight into surface processes that can be gained from the use of such potentials cannot be overestimated. The unknown accuracy of the model is a practical limitation but, as *ab initio* calculations on simulated surfaces containing a few surface atoms become more accurate, it will eventually be possible to monitor this.

6-8 Oxidation of silicon

A critical stage in the formation of a silicon chip is the formation of a layer of SiO_2 on its surface. This seals the inner surface of the Si from later contamination. The process occurs when the Si is strongly heated in the presence of oxygen. After the surface atoms have reacted, the layer enlarges by penetration of the oxygen into the Si. For further details see Grove [11].

The stages in this oxidation-diffusion process are:
1. The flow of gas to the surface
2. The penetration of the surface which is governed by Henry's law
3. The diffusion of the gas through the oxide layer to the Si surface
4. The reaction of the Si atoms with the oxygen.

The concentration of oxygen at various points is needed. At the outside surface this concentration is lowered because of the penetration so there is a flow towards the surface given by

$$F_1 = h_g (C_g - C_s)$$

where C_g, C_s are the concentrations far from the surface and at the surface. By Henry's law the concentration just inside the surface is proportional to that outside so this will be

$$C_0 = pC_s.$$

The flow across the oxide layer depends on the thickness, x, of that layer and is

$$F_2 = D(C_0 - C_i)/x$$

where C_i is the concentration at the start of the Si solid and D the diffusion constant. The reaction rate will depend on this concentration so that the flow of new oxide will be

$$F_3 = kC_i.$$

The oxide layer will grow at a rate given by

$$F = N\frac{dx}{dt} = \frac{kC^*}{1+k/h+kx/D}$$

where N is the number of oxide molecules per unit volume, $h=ph_g$ and $C^* = pC_s$. This differential equation can be solved and the solution has the form

$$x = A\sqrt{(1 + B[t+\tau])} -1$$

where $A=D(1/k + 1/h)$, $B=2DC^*/(NA^2)$ and τ depends on the initial thickness of the layer, with $\tau=0$ if this vanishes. The experimental results follow closely this model.

The details of this process show that it is possible to use classical macroscopic equations to arrive at an understanding of a complicated reaction.

6-9 The absorption of hydrogen in palladium

The process of hydrogen meeting the surface of a palladium crystal has similarities with that of oxygen on silicon. Initially the molecules of hydrogen are adsorbed on the surface as atoms. The study of the adsorption indicates that the dissociation of the molecule is the rate-determining step. The situation changes as soon as a large fraction of the surface has been covered. It then is found that some atoms can penetrate the surface and go into interstitial positions in lower layers of the crystal. Some experiments (Eley and Pearson, [12]) indicate that the critical fraction of coverage is 75%. At this point the heat of absorption equals the heat of adsorption.

6-10 Penrose tiling

The only rotations about a point which are compatible with a translation lattice are C_2, C_3, C_4 and C_6. In particular a five-fold rotation is not allowed. This constraint has important consequences when we consider the packing of molecules that have five-fold coordination.

A new possibility of building up a two-dimensional structure using elements which are compatible with five-fold symmetry has been suggested by Penrose [13]. He uses four kinds of basic tile. There is a pentagon and a five-pointed star. Another is a thin lozenge and the last is a three-pointed half star (or cocked hat). By placing these together, according to a systematic rule, he can cover the surface of a plane completely. There is exact five-fold symmetry around just one point. A selected portion, which uses just the upward pentagon, is shown in Figure 6. The five-pointed star and some downward pentagons emerge from the gap between pentagons. The other shapes are partly visible and become wholly so when as many downward pentagons as possible are added to the figure.

The result of the rule for placing the tiles is that the shapes around the centre are repeated outwards in a larger scale but with more elaborate boundaries. Thus the innermost shape of a pentagon surrounded by pentagons is repeated as shown by the added lines. It is also repeated on a larger scale to fill the figure. With care, the repeats which are its inversion can also be seen. Although this pattern does not repeat by

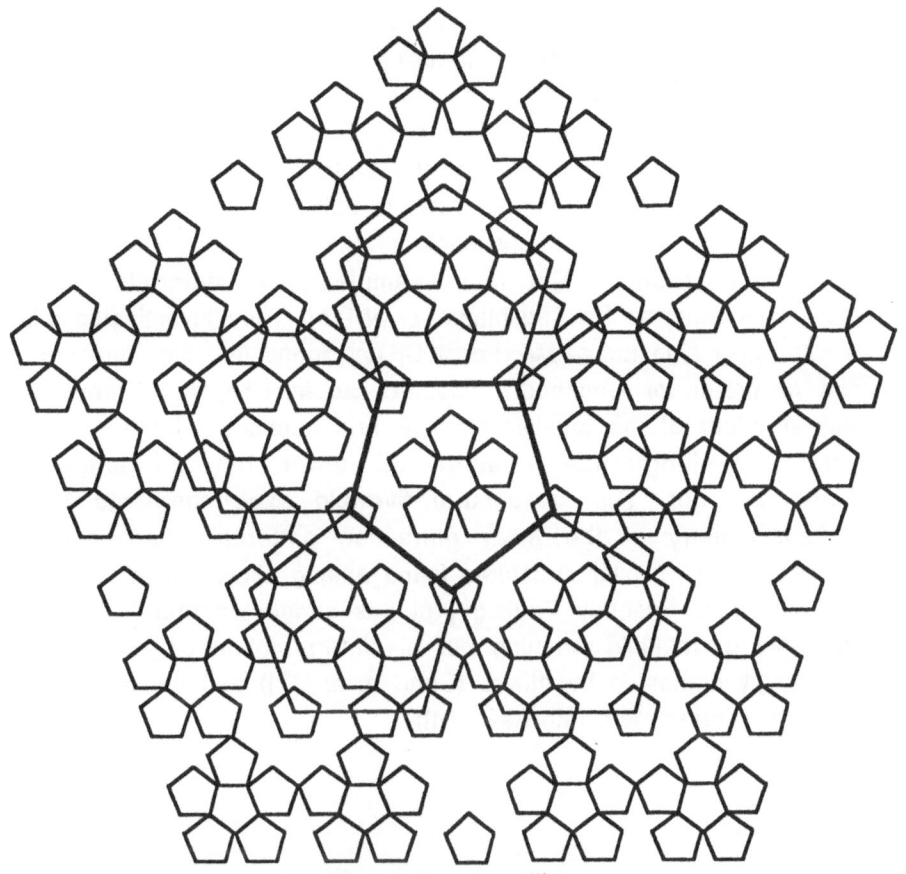

Fig. 6 Penrose tiling

translation on the same scale this local inner pattern does occur frequently. What has been drawn in Figure 6 consists very largely of its repeats. It is an example of another kind of relationship between short range order and long range order. Although this tiling does not have translation symmetry in two-dimensions it can be considered as a two-dimensional cross-section of a regular "cubic" lattice in five dimensions. A structure, such as this tiling, whose symmetry can only be understood as a section of a higher dimensional lattice is now called a quasi-crystal.

The demonstration that a complete covering of the plane is possible using tiles, which embody five-fold symmetry, has made many people think again about the possibility of new types of three-dimensional structure. The three-dimensional quasi-crystal (Mackay [14]), which generalizes the Penrose example to three dimensions, is irregular when considered as a crystal since it uses two rhombohedral unit cells systematically but it produces no long-range order unless it is seen as a projection from six dimensions. Yet the cells do conform to the local symmetry so that the short-range order is satisfied. The implication that short-range order does not necessarily produce periodicity is an insight of general importance. The first example of a material with this structure was reported by Shechtman et al [15]. They investigated alloys of Al with small amounts of Mn, Fe or Cr and found that they appeared to consist of grains of icosahedra (clusters as in Chapter 5) each built around an atom of the smaller element and connected by the Al. The material has long-range orientational order but no long-range translational order. The icosahedral five-fold axes contribute to the five-fold symmetry of the X-ray diffraction patterns. There are now many examples of similar structures found among alloys which have been rapidly solidified. Recently Pauling [16] has suggested that this alloy is an example of multiple twinning but his interpretation of the data has been strongly contested by others (Cahn et al [17]) and the quasi-crystal interpretation reaffirmed (Mackay [18]).

References

[1] Somorjai G A 1972 Principles of Surface Chemistry, Prentice-Hall Englewood Cliffs

[2] Lennard-Jones J E 1932 Trans Faraday Soc 28: 333

[3] Bardeen J 1940 Phys Rev 58: 727

[4] Mavroyannis C 1963 Mol Phys 6: 593

[5] Suthers R A, Linnett J W and Erickson W D 1974 Surface and Defect Properties of Solids 3: 132

[6] Shoenberg D 1952 Phil Trans Roy Soc A245: 1

[7] Johnston D F 1956 Proc Roy Soc A237: 48

[8] Pan R P, Etters R D and Kobashi K 1982 J Chem Phys 77: 1035

[9] Watanabe N, Nakajima T and Touhara H 1988 Graphite fluorides, Elsevier Amsterdam

[10] McCreery J H and Wolken G 1975 J Chem Phys 63: 2340

[11] Grove A S 1967 Physics and Technology of Semiconductor Devices, Wiley New York

[12] Eley D D and Pearson E J 1978 J Chem Soc Faraday I 74: 223

[13] Penrose, R 1974 Bull Inst Math Appl 10: 266

[14] Mackay, A L 1982 Physica A114: 609

[15] Shechtman, D Blech, I Gratias, D and Cahn, J W 1984 Phys Rev Lett 52: 1951

[16] Pauling L 1985 Nature 317: 512

[17] Cahn J W, Gratias D, Shechtman D 1986 Nature 319: 102

[18] Mackay A L 1986 Nature 319: 103

Chapter 7   Cooperative effects

7-1 Introduction

There are several properties of crystals which can exhibit cooperative effects under certain circumstances. Ferromagnetism is one familiar example, where the magnetic moment in one cell helps to induce a parallel moment in its neighbours and, hence, throughout the crystal. Ferroelectricity has a similar origin in the local dipole moments. It is less obvious that the structural properties of alloys are another example but they can also be included.

In this Chapter we will look at a very simple theoretical model for such effects (based on the work of Bragg and Williams, [1], on alloys) to show something of their nature. The limitations of the model will be pointed out in each example to which it is applied. Inevitably, the model will be mathematical.

7-2 The model

We consider an alternant crystal which can be described as one composed of two similar interpenetrating lattices such that all the vertices which are nearest neighbours to the vertices of one belong to the other. The body-centred crystal is an easy example since it is composed of two lattices of which one, the A lattice, is simple cubic and the second, the B lattice, is also simple cubic but with its vertices at the centres of the cubes of the first. Let there be N atoms on the A lattice and N on B. A typical vertex of the first lattice will be called a and b will be one of the second. The number of nearest neighbours of any vertex will be z. For the body-centred cubic lattice z=8.

We also assume that the crystal is composed of two kinds of atom so that, in the perfect crystal, each *a* site has an α atom and each *b* site has a β atom. We assume that only nearest neighbours interact in the sense of contributing to the total energy. This total, for the fully-ordered crystal, is

$$E_0 = Nz\, E_{\alpha\beta}$$

where $E_{\alpha\beta}$ is the interaction energy of two nearest neighbours, an α on A and a β on B. If there is an exchange so that one A site has a β atom and a neighbouring site of B has the α atom then there is a change of energy that now involves $E_{\alpha\alpha}$ and $E_{\beta\beta}$, the energies of nearest neighbours that are both α atoms or both β atoms. Thus, if there is one interchange of nearest neighbours, the increase in total energy will be

$$\Delta E = (z-1) \left(E_{\alpha\alpha} + E_{\beta\beta} - 2E_{\alpha\beta} \right).$$

If the site which has the β atom is a non-neighbour site then

$$\Delta E = z \left(E_{\alpha\alpha} + E_{\beta\beta} - 2E_{\alpha\beta} \right).$$

We now introduce a measure of the extent of these interchanges. Let n be the number of α atoms on the B vertices so that (N-n) remain on *a* sites. Similarly, for the β atoms, there must now be n on the A lattice to replace those removed and (N-n) on *b* sites. We define the order parameter λ to be

$$\lambda = 1 - 2n/N$$

so that $\lambda=1$ corresponds to perfect ordering (i.e. n=0) and $\lambda=-1$ corresponds to all the α atoms being on B (n=N), which can also be considered as a perfect ordering. $\lambda=0$ means that atoms are as likely to be on one lattice as on the other (n=N/2), which is a random allocation. Thus we have

$$n = N(1-\lambda)/2, \quad N-n = N(1+\lambda)/2.$$

The model now assumes that the crystal is homogeneous. This means that around each site of A there will be, on average, a mixture of atoms as determined by the macroscopic ratios of the atoms on B. Thus there would be zn/N that were α atoms moved from A and z(N-n)/N that were β atoms. Now (N-n) of these sites of A have α atoms so their contribution to the energy will be

$$(N-n)\left(zn/N\, E_{\alpha\alpha} + z(N-n)/N\, E_{\alpha\beta} \right)$$

while those with β atoms contribute

$n\left(zn/N\ E_{\alpha\beta} + z(N-n)/N\ E_{\beta\beta}\right).$

The change in total energy is found by summing over all the nearest neighbours of each site of A and then over all the sites of A. This gives, using E_0 above,

$\Delta E = zn(N-n)/N\ E_{\alpha\alpha} + zn(N-n)/N\ E_{\beta\beta} + z\left([(N-n)^2 + n^2]/N - N\right) E_{\alpha\beta}$

It is convenient to define J as

$J = E_{\alpha\alpha} + E_{\beta\beta} - 2E_{\alpha\beta}$

so that

$\Delta E = n(N-n)/N\ zJ$

$\qquad = N(1-\lambda^2)zJ/4.$

There is also a change in entropy. The α atoms can be distributed with (N-n) on one lattice and n on the other in $N!/(N-n)!\ n!$ ways. The β atoms must then occupy the vacant sites in the same number of ways so that the entropy change (as a pure number) is

$\Delta S = 2(\ln N! - \ln (N-n)! - \ln n!)$

and, using the familiar Stirling approximation

$n! = \left(\dfrac{n}{e}\right)^n,$

this reduces to

$\Delta S = 2\left(N \ln \dfrac{N}{e} - (N-n) \ln \dfrac{N-n}{e} - n \ln \dfrac{n}{e}\right)$

$\qquad = N\{2\ln 2 - (1+\lambda) \ln (1+\lambda) - (1-\lambda) \ln (1-\lambda)\}.$

To determine the value of λ when the system is in equilibrium at temperature T we have to minimize the free energy with respect to λ. Since the free energy change is

$\Delta F = \Delta E - kT\Delta S$

its differentiation shows that λ is the value which satisfies the equation

$\lambda = \text{th} \left(\dfrac{\lambda zJ}{4kT}\right).$

The non-zero solution of this equation for λ can be found from tables when the value of zJ/4k is given. It results in a value which starts at 1 when T=0 and reduces to 0 at a critical temperature T_c=zJ/4k. Thus, if heteronuclear binding is strongly favoured (J large), the ordered state persists to a high temperature before the thermal motion destroys it. For $T > T_c$ the only solution is λ=0 i.e. a random distribution. In a corresponding way the energy E increases with temperature through the critical point by NzJ/4 and the entropy by 2Nln2. The specific heat at constant volume has a finite discontinuity of 3k/2 at T_C.

7-3 The square lattice

There is an immediate check on the accuracy of this model. The two-dimensional square lattice with atoms at each vertex and nearest neighbour interactions has been extensively studied and has an exact solution (Onsager, [2]). This involves more sophisticated matrix techniques.

According to the model above the transition temperature is at $T_C=zJ/4k$. The square lattice has z=4 so $T_C=J/k$. The exact solution has $T_C=0.569J/k$. The total increase in E was NzJ/4 i.e. NJ and this is also the exact value. The entropy change is also the same. It corresponds to the loss of information (ln2) on the identity of the atom occupying each of the 2N sites. The specific heat had a discontinuity in the model at T_C but the exact solution has a logarithmic singularity there. Thus some features of the exact solution are given well by the model but others are not very accurate. The weak point in the model is its assumption that the composition of the local environment can be identified with its bulk behaviour. More accurate models refine the description of this short range order.

7-4 Alloys

An alloy such as CuAu or CuZn (β brass) is a good example of a system which this theory tries to treat. At low temperatures the lattice is ordered with one kind of atom on one of the alternant lattices and the other kind on the other. The arrangement becomes less ordered as the temperature rises until there is a critical temperature above which the arrangement is random. There is a discontinuity in the specific heat at this critical temperature. Thus the qualitative picture is correct.

When we compare quantitative predictions the model is not so effective. The shape of the specific heat around the critical point is different in several respects from experiment. In particular, experiment shows some increase persisting on the upper side of the critical temperature and the model does not. This has led to the refinement of the theory. In one respect the theory is obviously inadequate. Its assumption of homogeneity, that the local ordering is the same as the macroscopic ordering, is incorrect. What we have measured is the long-range order. What we need is the short-range order. This becomes clear if we take the extreme example of an arrangement in which half the crystal is correctly ordered ($\lambda=1$) and the adjacent half is exactly wrong ($\lambda=-1$). The combined value of $\lambda=0$ but, in fact, only the sites at the junction of the halves are incorrect! The theories which try to calculate the short-range order more exactly are too complicated to be described here.

7-5 Melting

A direct use of this model of cooperative effects was made by Lennard-Jones and Devonshire [3] in their discussion of melting. They visualised an atomic crystal with, around each lattice site, another set of interstitial positions which are unoccupied at low temperatures but can become thermally occupied as the temperature rises. In the liquid state both sites would be occupied at random. Thus, in this application, the B lattice is initially empty and its atoms are "holes". The transition corresponds to melting and the critical value of λ determines the melting temperature exactly as before.

It is important in this situation to consider the volume dependence of the free energy. The energy difference between the normal site and the interstitial site is due to the repulsive potentials of the atoms around the interstitial site. The repulsive potential will depend on the inverse 12th power of the distance and, hence, to the inverse 4th power of the volume. The free volume available to the atom increases with temperature because of the increased amplitude of the thermal vibrations. This can be evaluated using the interatomic potential. Thus the energy difference is large at low temperatures and forces the state to be ordered but it decreases at higher temperatures and permits the onset of disorder.

When related to the Ar crystal this theory gives for the volume increase at melting 13.5% which compares with 12% experimentally. For the change of entropy on melting this gives 1.7 as compared with 1.66 from experiment (see Chapter 1).

7-6 Nematic liquid crystals

The nematic phase of a liquid crystal is one in which the long-chain molecules still have their axes largely pointing in the same direction but have their other degrees of freedom randomized. This means that they have rotational freedom and their centres of mass are not aligned. A general introduction to the variety of liquid crystals and their properties has been given by Priestley et al [4].

The intermolecular potential is due to the atom-atom potentials we have already discussed but to achieve a simpler theory we consider a single potential from one molecule to the other. The important variable is θ, the orientation of the molecular axis with respect to the direction of the crystal (usually called the director). We are not usually interested in whether the head or the tail points in this direction so the variable used to describe the situation should reflect this. The practical choice is to use the value of $P_2(\cos\theta)=(3\cos^2\theta-1)/2$ as the order variable. This has the value 1 when the axis is parallel to the director ($\theta=0$ or $\theta=\pi$) and the value 0 when the axis is in a random direction ($<\cos^2\theta>=1/3$).

The potential of two neighbouring molecules depends on θ_{12}, the orientation of one axis with respect to the other. The representation of this is

$U = C\, P_2(\cos\theta_{12})$

$= C\{ P_2(\cos\theta_1)\, P_2(\cos\theta_2) + 3\sin^2\theta_1\,\sin^2\theta_2\,\cos(\phi_1 - \phi_2)\}$

where C is a constant because all other variables have been eliminated. The term in $(\phi_1-\phi_2)$ will vanish when the average is taken over these variables. These contributions are to be summed over all neighbours. The total potential, when the attractive nature of the dispersion forces is considered, can be written as

$V = -c\, P_2(\cos\theta_1)\, P_2(\cos\theta_2),\quad c>0.$

When the second molecule is averaged this becomes

$V = -c\, P_2(\cos\theta_2)\, P$

where $P = \langle P_2\rangle$. The Boltzmann equation shows how this average is found:

$$\langle P_2\rangle = \int_0^1 P_2(\cos\theta)\, e^{cP_2(\cos\theta)P\beta}\,d(\cos\theta) \Big/ \int_0^1 e^{cP_2(\cos\theta)P\beta}\,d(\cos\theta).$$

This is more easily understood in terms of the partition function

$$Z(\beta) = \int_0^1 e^{cP_2(\cos\theta)P\beta}\,d(\cos\theta)$$

from which the mean value is given by

$\langle P_2\rangle = 1/(c\beta Z)\,\partial Z/\partial P$

Since P appears inside the integral this is an implicit equation for P as a function of β ($\beta=1/kT$). The equation can be solved by a numerical integration or by various expansions of the integrals. The results are shown in Figure 1.

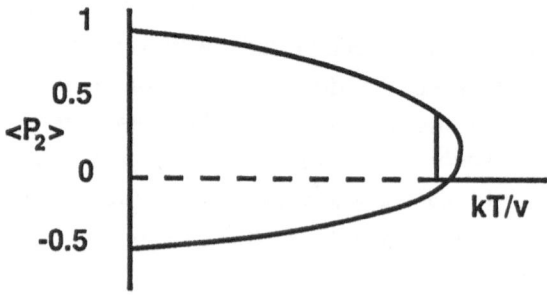

Fig.1 The order parameter P as a function of kT/v.

There are three branches for P in the neighbourhood of the origin of $1/\beta$ i.e. for small temperatures. The stable solution starts from the value 1 showing complete ordering of the axes. It decreases to a value of 0.4289, at the critical temperature of T=0.2202c/k, where the next branch, having P=0, becomes more stable. There is, therefore, a discontinuity in the value of P at this point.

From this the thermodynamic functions can be deduced. They give the entropy change at the transition as
$\Delta S = 0.418$.
This shows the small amount of order remaining in the nematic phase.

7-7 The smectic-A phase of liquid crystals

In the smectic phases of a liquid crystal there is rather more order than in the nematic phase. One of these, smectic-A, has all the molecules with their centres of mass in planes normal to the director. This is equivalent to having their heads on a plane and their tails on a parallel plane. In the smectic-C phase the director is at an angle to these parallel planes. Thus, in addition to the alignment of the axes, there is now a periodicity of the molecules in the director direction.

A theory of this phase can be set up using the same ideas as before. The intermolecular potential must now contain a variable to express the periodicity of the position of the centre of mass in the director direction. An appropriate variable is the first of the Fourier coefficients $< \cos(2\pi z/d)>$, where d is the distance between layers of the molecules and z is measured along the director. By the same form of argument as before, we take the average of the neighbouring molecules assuming that they have the same distribution as the bulk material. The result is an effective one-molecule potential:

$$V = -v\left(\delta\alpha\tau \cos\left(\frac{2\pi z}{d}\right) + \left(\eta + \alpha\sigma \cos\left(\frac{2\pi z}{d}\right)\right)P_2(\cos\theta)\right)$$

where the phase is now described by the three order parameters
$\eta = <P_2>$
$\tau = < \cos(2\pi z/d)>$
$\sigma = < P_2(\cos\theta) \cos(2\pi z/d)>$

and the constants v, α, δ allow the three terms to have different values. The single molecule partition function is:

$$Z(\beta) = \int_0^1 \int_0^d e^{-\beta V} dz \, d(\cos\theta)$$

From this, the mean values can be determined as
$\eta = 1/(\beta v Z) \, \partial Z/\partial\eta$, $\tau = 1/(\beta v \delta \alpha Z) \, \partial Z/\partial\tau$, $\sigma = 1/(\beta v \alpha Z) \, \partial Z/\partial\sigma$.

These implicit equations determine the parameters as functions of temperature through $\beta = 1/kT$. As before, these cannot be solved in closed form and numerical techniques have to be used. It is necessary to fix some of the variables in order to do so. Typical results show transition points between the low-temperature smectic phase into the nematic phase and from it into the liquid phase itself. Although these models of liquid crystals have introduced approximations into the potential on top of the approximations of the mean field model they do serve a useful purpose in describing the major features of their behaviour.

7-8 Ferromagnets

An atom which has a magnetic moment μ located in its inner shell and free to rotate in an applied field will have an additional energy $-\mu.H$ in the presence of a magnetic field. We consider an elementary argument using the mean field concept. The related factor in the partition function will be:

$$P(T) = \int e^{(\mu.H)/kT} \, d\omega$$

where ω is the element of solid angle. It is convenient to take a coordinate system whose z coordinate is in the direction of the magnetic field. This reduces the integral to:

$$P(T) = 2\pi \int_{-1}^{1} e^{\mu H c/kT} \, dc$$

$$= \frac{4\pi kT}{\mu H} \sinh\left(\frac{\mu H}{kT}\right)$$

The average value of the moment in the direction of the field will be

$$<\mu> = kT \, \partial \ln P / \partial H = \mu \left(\coth \frac{\mu H}{kT} - \frac{k \, T}{\mu H} \right)$$

and, for most external fields $\mu H \ll kT$, so that this can be approximated as

$$<\mu> = \frac{\mu^2 H}{3kT}.$$

The susceptibility is found as the ratio

$$\chi = N \left(\frac{<\mu>}{H} \right) = \frac{N\mu^2}{3kT}.$$

This is the usual Curie law for paramagnetic material. For ferromagnetics we have to assume that the atom experiences another magnetic field due to its neighbours. This Weiss field will be assumed to be proportional to the average magnetic moments of these neighbours. In this treatment the origin of this term is not explained. The coupling constant is too large to be due to any direct interaction of the neighbouring moments. The usual explanation involves the exchange forces between the spin moments. A further discussion, using the Hubbard model, is given in Appendix 5. As above, this mean field argument assumes the homogeneity of the material. If the Weiss field is:

$$H_W = \lambda N <\mu>$$

then the induced moment becomes:

$$N<\mu> = \frac{N\mu^2}{3kT} (H + \lambda N <\mu>).$$

The magnetic susceptibility is now

$$\chi = \frac{N\mu^2}{3k(T - \theta)}, \, \theta = \frac{\lambda N \mu^2}{3k}$$

This is the Curie-Weiss law and is a fair representation of the susceptibility above the critical temperature θ. Below this temperature the internal field is strong so the full form for $<\mu>$ is needed. It can then be seen that $<\mu>$ will not vanish when the external field is removed so that the material remains magnetized. More accurate expressions for the magnetization depend on using the spatial quantization of the spin magnet moments in the presence of a field instead of the classical averaging over all angles.

7-9 Ferroelectrics

There are many molecules which possess a dipole moment and so give rise to dipolar interaction throughout the crystal. Many crystals cannot produce any cooperative effect because the molecule in the crystal is in a symmetrical position so that the interaction with one neighbour cancels the interaction of another. If the site symmetry is at least octahedral then the sum of the contributions from successive shells of neighbours is zero. The interesting crystals are those which have reduced site symmetry.

The theory of ferroelectrics has been modelled on that of ferromagnetics. In particular the analogy of the Weiss field is required. This cannot be due to an exchange term since the molecules do not possess any spin moment. The Lorenz field is a possible mechanism but its magnitude is relatively small. It seems that a distortion of the lattice by the electric field, with loss of the local symmetry, is required to produce significant effects. This can remove the molecular dipole from a zero Ewald-field position to one which is quite intense. Since some distortions in certain crystals can produce a local electric field in the opposite direction to the source field antiferroelectrics are also possible.

7-10 Ferroelasticity

Some alloys, at low temperatures, have a martensite phase. This gives rise to the interesting phenomenon of shape memory. Industrial uses for these materials continue to be found (see, for example, the review by Schetky [5]).

At high temperatures the alloys of Ni and Ti developed by the US naval research have an austenite phase. This is a disordered bcc structure with the atoms of both species placed at random on the lattice points. Below 60°C the martensite phase is formed. The exact form of this depends on the alloy and the external pressure. Typically it is less symmetrical (orthorhombic instead of cubic) and allows one of the atomic species to move to preferred positions. In this form the material can easily be distorted by applied stress. It reacts by changing its structure from one form to another and not by the slipping of successive planes of the material against one another. For this reason the material can recover its original configuration by raising its temperature above the transition temperature to enable the atoms to pass over the barrier into their more stable martensite configuration.

It is more difficult to relate the properties of this material to the theory we have been developing since no one variable describes the transition. Both geometrical distortions and local phase changes are involved. The analogy with ferromagnetics is closest since the material shows such properties as hysteresis under stress and storage of energy.

References

[1] Bragg W L and Williams E J 1935 Proc Roy Soc A152: 231
[2] Onsager L 1944 Phys Rev 65: 117
[3] Lennard-Jones J E and Devonshire A F 1939 Proc Roy Soc A170: 464
[4] Priestley E B, Wojtowicz P J and Sheng P 1975 Introduction to liquid crystals, Plenum New York and London
[5] Schetky L M 1979 Sci Amer 241 (5): 6874

Appendix 1 Lattice sums

A1-1 Linear sums

The energy of interaction between two charges is the product of their charges divided by the distance between them. When there is an infinite set of charges equally spaced along a line, each of the same magnitude but alternating in sign, their energy is a conditionally convergent sum and its evaluation requires great care. It is a well-known theorem that an method of summation which modifies the order of the terms can arrive at a completely arbitrary result.

The formal expression for the energy of one charge in this linear array, when we select as the unit of charge the magnitude of each charge, as the unit of distance the nearest neighbour distance and as the unit of energy this charge squared divided by this distance, is written as -2E, where

$$E = 1 - \frac{1}{2} + \frac{1}{3} - \frac{1}{4} \ldots = \sum_n \frac{(-1)^n}{n}$$

This is a well-known series whose sum is ln(2)=0.693147. The partial sums of the series alternate on either side of this but the terms decrease so slowly that this direct method of calculation is inefficient. To obtain six digit accuracy requires at least one million terms!

A1-2 The supercell method

A better method (based on that by Evjen [1]) of performing this linear sum is to superimpose a periodicity on the line with identical supercells each of which is neutral and has no dipole moment. They are centred on the positive charges. The contribution of the central supercell is then found explicitly from the finite sum inside that cell while those of the more distant cells are found by first combining the contributions of those within the same supercell. The smallest supercell for the linear problem contains the central ion with its two first neighbours except that these charges are divided equally between the two supercells meeting there so that each supercell is identical and neutral. This gives the arrangement:

$$-\frac{1}{2} \quad 1 \quad -\frac{1}{2} \qquad\qquad -\frac{1}{2} \quad 1 \quad -\frac{1}{2} \qquad\qquad -\frac{1}{2} \quad 1 \quad -\frac{1}{2}$$
$$\qquad -\frac{1}{2} \quad 1 \quad -\frac{1}{2} \qquad\qquad -\frac{1}{2} \quad 1 \quad -\frac{1}{2}$$

In this, alternate supercells have been placed above and below to avoid confusion. The supercell length is twice the distance between nearest neighbours. The central supercell gives a contribution of $-\frac{1}{2}-\frac{1}{2}=-1$. The supercells on either side of it give smaller contributions because they are further apart. The nth supercell is easily shown to contribute $-\frac{1}{n(4n^2-1)}$. These terms still add to give $-2E$ so that E is estimated to be:

$$E = \frac{1}{2} + \frac{1}{6} + \frac{1}{60} + \frac{1}{210} + \frac{1}{504} + \frac{1}{990} + ... = 0.691089$$

Thus we have transformed the original series into a series which has positive terms and converges fairly quickly. By using supercells of twice this size and including some of the second neighbour charge it is possible to produce a set of overlapping supercells, still centred on the positive charges, so that each supercell has no total charge, no dipole and no quadrupole moment. The series so obtained converges even faster:

$$E = \left(1 - \frac{7}{16} + \frac{1}{6} - \frac{1}{32}\right) - \sum_{2} \frac{3}{8n(n^2-1)(4n^2-1)}$$

This gives, after only three terms of the summation, $E=0.693204$. This series has negative terms, after the contributions of the central supercell, and so approaches the limit from above whereas the earlier series approached from below. Together the two series give an estimate of the error. It is readily observed that the supercell sum includes the entire contributions from the ions close to the original ion but the boundary between those included in the partial sum and those neglected has been made broader and less abrupt.

A1-3 Sums in three dimensions

When we consider lattice sums in three dimensions we have additional problems. Along a line the sequence of the contributions is fixed but in space even the ordering of the terms has to be determined. It is tempting to put together all the terms at the same distance from the chosen centre. The number of neighbours of different orders increases approximately as the square of the distance whereas the potential decreases as the inverse distance so that the resulting series has terms which increase in magnitude and the partial sums oscillate divergently. The NaCl lattice, using Table 2.1, gives

$$\alpha = 6 - \frac{12}{\sqrt{2}} + \frac{8}{\sqrt{3}} - \frac{6}{2} + ...$$

It is easily seen that these leading terms give wildly divergent partial sums and do not approach the proper limit which is the Madelung constant $\alpha = 1.74756$! This has proved not to be a practical procedure.

The device suggested above is of some help. The cubic lattice, which consists of all the ion sites in the NaCl lattice, can be used to define a set of supercells. By selecting a supercell which is twice the size of the original cube in each direction and so contains eight cubes a supercell is obtained which has a centre of symmetry and so has no dipole moment. It would have a net charge so the ions on its boundary have to be divided as before. An ion in a face is equally divided between the supercell and its neighbour. An ion at a vertex belongs to eight supercells and so is equally divided between them. An edge ion is divided between four supercells. The finite sum over the central supercell alone now gives the estimate:

$\alpha = 1.45603$

and including the six nearest-neighbour supercells brings this to the value:

$\alpha = 1.86968$

With the twelve supercells that come next the sum becomes

$\alpha = 1.71926$

Thus the approximation to the correct sum, 1.74756, becomes apparent. This approach can be extended to other lattices and will produce a slowly convergent sum. The use of larger, overlapping supercells which minimize the quadrupole components would produce a more rapidly converging series.

A1-4 The Ewald method

To obtain a fast convergence of the lattice sums and to deal with less symmetrical lattices some new principles are required. Ewald [2] has proved that the use of three-dimensional Fourier series and certain properties of theta functions can be a most efficient method. The simplest theta function can be defined as the Fourier series:

$$\theta(x,q) = \sum_{n=-\infty}^{\infty} q^{n^2} e^{2nix}$$

He considers the potential due to a set of three-dimensional Gaussian functions located at the ionic nuclei and with the same total charges. Since the potential of the Gaussian has the same spherical moments as that of the delta function these have the same asymptotic behaviour. The exact potential can then be expressed as the potential of the Gaussians along with a term to correct for the difference. Since the potential of the Gaussians can be expressed using the error function and other known functions the problem can be reduced to the evaluation of two contributions which are each rapidly convergent series. A very readable account of this argument has been given by Ziman [3]. Computer programs to perform these summations are now available.

References

[1] Evjen H M 1932 Phys Rev 39:675

[2] Ewald P P 1917 Ann d Phys 54:519, 1921 Ibid 64:253

[3] Ziman J M 1964 Principles in the Theory of Solids, Cambridge London

Appendix 2 The phase difference method

A2-1 Introduction

In this book we find, in several different contexts, the problem of diagonalizing a matrix of circulant type. This has, in its simplest form, vanishing elements except for those on the diagonal itself, which are all equal to α, and those immediately above and below the main diagonal, which are all equal to β. Such matrices arise when a structure consists of exactly similar units connected together in a linear form. These forms readily give rise to waves and the solution of the eigenvector problem uses this fact.

A2-2 Constant phase differences

The eigenvector problem for a circulant has the form:
$$\beta\, a_{n-1} + \alpha\, a_n + \beta\, a_{n+1} = \varepsilon\, a_n$$
where n numbers the various units, the eigenvector has the component a_n at unit n and ε is the eigenvalue. There is one similar equation for each unit. To solve this difference equation we put:
$$a_n = c e^{in\theta},$$
where c is a constant, and insert into the equation. This will be a solution for all values of n provided that:
$$\varepsilon = \beta e^{-i\theta} + \alpha + \beta e^{i\theta}$$
$$= \alpha + 2\beta \cos\theta$$
This is clearly the condition that determines the eigenvalue. The angle θ is a constant phase difference from one unit to the next. To avoid repeating the same solution it must lies between the values 0 and 2π.

This solution assumes that the system is infinite. To become more realistic for a molecular system we assume that the system has a finite number of identical units and that these are arranged in a circle so that the Nth unit is also the zero unit. This implies that the components satisfy

$$a_N = a_0$$

so

$$e^{iN\theta} = 1$$

which has the solutions

$$\theta = 0, \frac{2\pi}{N}, 2\frac{2\pi}{N}, \ldots , (N-1)\frac{2\pi}{N}$$

There are N of these, so this is the full set of solutions. It is important to note that they are equally spaced in the variable θ. The eigenvectors are normalized by putting $c=\sqrt{\frac{1}{N}}$ and the eigenvalues become:

$$\varepsilon(r) = \alpha + 2\beta \cos\frac{2r\pi}{N}, \quad r = 0, 1, 2, .. (N-1).$$

It should be noted that, because of the symmetry of the cosine function, the eigenvalues with r and with (N-r) are identical. These eigenvalues are doubly-degenerate while the r=0 eigenvalue is singly-degenerate. If N is even the r=N/2 eigenvalue is also singly-degenerate. The eigenvectors a_n corresponding to these describe waves which travel round the circle. Those with r and (N-r) are waves with the same wavelength but travelling in opposite directions.

If, instead of being in a circle, the system has finite length and then stops the solution can be adapted to meet this modification. The two degenerate waves can be superimposed to give a standing wave and this will have nodes. The solutions required are those whose nodes coincide with the missing units on either side of the system. Thus, if the system runs from 1 to N, we require the nodes to be $a_0=0$ and $a_{N+1}=0$. These lead to

$$a_r = c \sin\frac{r n \pi}{N+1}$$

for the nth eigenvector and its eigenvalue is

$$\varepsilon(n) = \alpha + 2\beta \cos\frac{n\pi}{N+1}, \quad n=1, 2, .. \ N.$$

Note that these also are equally spaced in the variable θ.

A2-3 Density of States

Since the states are equally spaced in θ it is possible to evaluate the density of states for this system. If $n(\theta)$ is the number of states with phase difference less than θ then it is linear in θ. The density of states is the number of states of given energy as a function of that energy. Thus it is

$$\rho = \frac{dn}{d\theta}\left(\frac{d\varepsilon}{d\theta}\right)^{-1}$$

$$= \frac{1}{2\pi\beta\sin\theta} = \frac{1}{\pi\sqrt{4\beta^2 - (\varepsilon-\alpha)^2}}$$

The density becomes infinite at both ends of the energy band and has a minimum at $\varepsilon=\alpha$ its centre.

A2-4 Generalizations

This solution can be extended to deal with certain more elaborate systems. In particular, systems that repeat in two or three dimensions can be treated in the same way. The eigenvectors are waves with constant phase differences in each direction. The eigenvalues are additive. To adapt to finite systems it is usual to impose the cyclic boundary conditions (known as the Born-von Kármán boundary conditions) which are the generalization of the circle condition above.

In the text the problem arises of a linear repeating system where the unit has several different components in each repeat. This is treated by a simple modification of the basis idea (See Chapter 3 for an example). The components within the nth unit are collected into a subvector a_n, their interactions become a matrix A and all the matrix elements connecting it with the next unit become a matrix B. The infinite circulant system, when partitioned in this way, has the form:

$$B^T a_{n-1} + A a_n + B a_{n+1} = \varepsilon a_n$$

where the dimensions of the subvectors and matrices equal the number of components within a unit and T is the transpose operator. The constant phase difference method suggests the substitution:

$a_n = ce^{i\theta}$.

This will be a solution provided that a_n is an eigenvector of the matrix

$(A + Be^{i\theta} + B^Te^{-i\theta})$

The corresponding eigenvalue equation is

$|(A-\varepsilon I) + Be^{i\theta} + B^Te^{-i\theta}| = 0$

where I is the unit matrix. With r components, this gives r solutions for the eigenvalues $\varepsilon(\theta)$ which means that there are r bands whose energies are functions of θ.

Another generalization which can be achieved immediately is the inclusion of second neighbour interactions. Thus, if the equation for the linear system is

$\gamma a_{n-2} + \beta a_{n-1} + \alpha a_n + \beta a_{n+1} + \gamma a_{n+2} = \varepsilon a_n$,

then the constant phase solution still holds

$a_n = ce^{i\theta}$

but the eigenvalue relation becomes

$\varepsilon = \alpha + 2\beta \cos \theta + 2\gamma \cos 2\theta$.

Thus the shape of the energy band is changed but the form of the eigenvector remains wave-like.

Appendix 3 Impurities in bands

A3-1 Introduction

We consider here a simple model of the problem of one impurity atom introduced into an infinite linear chain. As in Appendix 2, the Hückel model of the chain leads to the equations:

$$\beta a_{r-1} + \alpha a_r + \beta a_{r+1} = \varepsilon a_r, \quad r \neq 0.$$

but at r=0, the impurity position, the energy α is replaced by A, without changing the interaction β, so that the remaining equation is

$$\beta a_{-1} + A a_0 + \beta a_1 = \varepsilon a_0.$$

Away from the impurity the equations are the same as before so the same solution must hold there:

$$a_r = a e^{ir\theta}$$

which is the constant phase difference solution. Its energy is still

$$\varepsilon = \alpha + 2\beta \cos \theta.$$

It is easier to work with the standing wave solutions obtained by combining the degenerate travelling solutions. One of these will be anti-symmetric with respect to the origin at the impurity and this must be undisturbed by the impurity since it has zero amplitude there. Its form will be:

$$a_r = a \sin r\theta.$$

The symmetric solution has to be modified to fit the equation for a_0. It can be written as

$$a_r = a \cos(|r|\theta - \phi)$$

where the extra phase difference ϕ is adjusted to satisfy the r=0 equation:

$$\tan \phi = \frac{\alpha - A}{2\beta \sin \theta}$$

These two solutions remain degenerate in energy.

These solutions, for all θ, constitute a complete set of standing wave solutions. They can be modified, by addition or subtraction of the degenerate pairs, to recreate the travelling wave solutions but the phase difference at the origin produces an extra reflected wave which was not present in the pure chain.

A3-2 The extra state

Outside the band there is one extra solution which is localized. It has the form

$$a_r = a\, m^{|r|}$$

where m is determined by

$$m = \frac{\varepsilon - A}{2\beta}$$

The corresponding energy eigenvalue is

$$\varepsilon = \alpha \pm \sqrt{4\beta^2 + (\alpha - A)^2}$$

where the sign must be taken to ensure that $|m| < 1$. This condition ensures that the solution becomes exponentially small in both directions so that it is localized around the impurity. If A lies below α, the negative sign is taken and m is positive while, if A is above, the sign is positive and m is negative. Thus, for A above α the energy of this solution splits off from the top of the band whereas for A below α it splits off from the bottom. When β is small compared with $|A-\alpha|$ the energy of the solution approximates to the impurity energy $\varepsilon = A$ and the effect of β is to allow for the polarization of this energy caused by the chain.

A3-3 Effect on the band structure

The solutions given above have implications for the band structure when the system contains an impurity. The impurity will leave the band itself looking exactly the same since the phase difference, ϕ, at the impurity does not change the energies of these solutions. The extra solution splits off from the band above or below the band depending on the sign of A-α. Its asymptotic behaviour is to approximate to the value of A. Figure 1 shows the band of the orbital energies and the extra energy as a function of A around A = α.

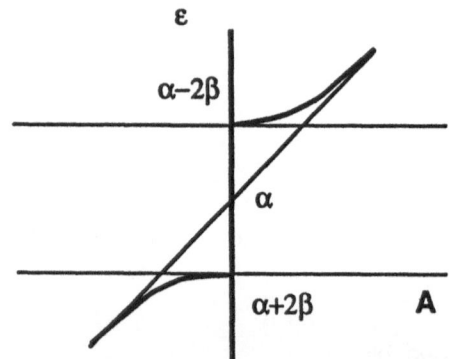

Fig. 1 Band structure with an impurity

Appendix 4 Second quantization

A4-1 Introduction

One of the barriers preventing a chemist from reading the modern literature of Solid State Physics is its use of second quantization notation and concepts. This appendix presents a very simplified introduction to this notation and to a few of its ideas. A more complete introduction has been given by Linderberg and Öhrn [1].

In order to relate to familiar situations we start from the translation of the Hückel theory into this notation.

A4-2 Hückel theory

A conjugated hydrocarbon with N carbon atoms has 2N π atomic orbitals (including spin) for its C atoms:

ω_r, $r=1,2...2N$.

It is important, for second quantization, to insist that these orbitals be orthonormal. Corresponding to each of these we define a creation operator:

$a_r^\dagger$,

which will insert the orbital into the total wavefunction, and an annihilation operator:

a_r,

which will remove it, if it is present. Thus, if we start from $|0>$, the vacuum wavefunction, which contains no electrons, the operators have the effect of building up a determinantal wavefunction as this example shows:

$a_1{}^\dagger a_2{}^\dagger \, |0> = a_1{}^\dagger \, |\omega_2(1)>$

$\qquad = \sqrt{\tfrac{1}{2}} \, |\omega_2(1)\omega_1(2) - \omega_2(2)\omega_1(1)>.$

This example also shows that the operators only act when they are immediately in front of the state and that they produce an antisymmetrical wavefunction which is normalized to 1. The operators do not commute but satisfy the relation

$a_r{}^\dagger a_s{}^\dagger + a_s{}^\dagger a_r{}^\dagger = 0$

so that an interchange of operators will produce a wavefunction with the opposite sign outside and with the two functions reversed inside.

The annihilation operators behave in a similar way. They will act on the wavefunction from the right to the left and will eliminate the orbital if it is present. If it is absent then the entire wavefunction is eliminated. These result in the corresponding relations:

$a_r a_s + a_s a_r = 0.$

It is to be noted that the squares of all these operators vanish. This is one illustration of the Pauli principle that no two electrons can have the identical orbital. When two different operators come together we have the third set of relations:

$a_r a_s{}^\dagger + a_s{}^\dagger a_r = \delta_{rs}$

With the use of these the action of any product of operators can be deduced. Another operator which is used frequently is the number operator:

$n_r = a_r{}^\dagger a_r$

This counts the number of times the orbital ω_r appears in the wavefunction. For example the combination $a_2{}^\dagger a_1{}^\dagger a_2$ has

$a_2{}^\dagger a_1{}^\dagger a_2 = -a_1{}^\dagger a_2{}^\dagger a_2 = -a_1{}^\dagger n_2.$

There is also the total number operator:

$N = \sum_r n_r$

which counts the total number of electrons irrespective of which orbitals they occupy.

With these operators we can construct the Hamiltonian. Its first term is the diagonal element representing the energy of an atomic orbital. Its value is α and it arises when the orbital is occupied so its form is:

$$\sum_r \alpha \, n_r = \sum_r \alpha \, a_r{}^\dagger a_r$$

where the sum is over all the possible orbitals (assuming all have equal integrals). The next term is the interaction of nearest neighbours. This has the form:

$$\sum_{(rs)} \beta \, a_r{}^\dagger a_s$$

where the sum is over all the orbitals r and over all the orbitals s which are nearest neighbours to them. The full Hamiltonian is:

$$H = \sum_r \alpha \, a_r{}^\dagger a_r + \sum_{(rs)} \beta \, a_r{}^\dagger a_s$$

The integral β is the resonance integral which plays a fundamental role in the Hückel theory. In the solid state literature it is called the hopping integral since it measures the rate of transfer of an electron from one orbital to the next.

The molecular orbitals are linear combinations of the atomic orbitals such that the Hückel matrix is diagonal with diagonal energies ε_r. The creation and annihilation operators can be similarly transformed linearly and the result will be the operators $c_r{}^\dagger$ and c_r. Using these, the Hamiltonian will take the diagonal form:

$$H = \sum_r \varepsilon_r \, c_r{}^\dagger c_r.$$

Reference

[1] Linderberg J and Öhrn Y 1973 Propagators in Quantum Chemistry, Academic Press London and New York

Appendix 5 Improved semi-empirical methods

A5-1 Introduction

Although the Hückel theory, and its solid state equivalent, is still a valuable model especially for large conjugated hydrocarbons it has been modified by the addition of extra terms which arise naturally from the self-consistent molecular orbital theory. These improve the agreement with experiment. The first of these (from 1953) comes from the work of Pariser, Parr and Pople and is usually called the PPP model. For a recent review and commentary see [1].

The PPP model is still a semi-empirical model and the agreement with experiment, for excited states in particular, is very sensitive to the values of some parameters. The implication is that the model already incorporates elements of a more accurate theory which has the same form of equation but modified values of the integrals. Some progress has been made in identifying this theory but the result is not yet clear.

In solid state theory a rather similar set of ideas for modifying the tight-binding model has been developed, apparently without any knowledge of the parallel developments in quantum chemistry. This started with the work of Anderson and the extensions made by Hubbard. These were designed originally to apply to the d bands of ferromagnetics though their general implications came to be appreciated.

These models have a valuable use in treating the surface states of solids and the process of chemisorption. In this context the use of Green's function methods (see Appendix 7) is found to be helpful.

A5-2 The PPP model

The PPP model is based on the exact equations [2] for the best molecular orbitals. These have the form:

$$F\psi = \varepsilon\psi$$

where F is the Fock operator, the effective one-electron Hamiltonian, and ψ is one of the molecular orbitals whose energy is ε. When the molecular orbitals are expanded in terms of atomic orbitals [3,4] this becomes a non-linear matrix equation. In the conventional derivation the Slater atomic orbitals are made orthogonal by the Löwdin technique [5]. This orthogonalization has the effect of producing an immense simplification of the two-electron integrals so that only those of Coulomb type need be included in the expressions. An alternative derivation [6] uses a highly-excited ferromagnetic state of the molecule to define the localized atomic orbitals as already orthogonal. This has the advantage that these orbitals can be optimized in relation to the molecule. The final result is very similar.

If the orthogonal atomic orbitals are ω_s and the Coulomb integrals are:

$$\gamma_{st} = \int \omega_s^*(1)\omega_s(1) \frac{1}{r_{12}} \omega_t^*(2)\omega_t(2) \, d\tau_{12}$$

then the diagonal elements of the Fock matrix are

$$F_{ss} = U_{ss} + 1/2 \, P_{ss}\gamma_{ss} + \sum_{u \neq s}(P_{uu} - 1)\gamma_{us}$$

where the first term contains the integrals over the one-electron operators and the σ core, the second allows for the extra repulsion when the atomic orbital is doubly-occupied and the third is the Coulomb contribution from its neighbours. P_{uu} is the electronic charge on atom u.

The off-diagonal elements are:

$$F_{st} = \beta_{st} - 1/2 P_{st} \gamma_{st}$$

where P_{st} is the bond order matrix. If the molecular orbitals are

$$\psi_s = \sum_u \omega_u a_{us}$$

then the bond order matrix, for doubly-occupied orbitals, is:

$$P_{uv} = 2 \sum_{s \, occ} a_{us}^* a_{vs}$$

In practice, for conjugated hydrocarbons, the core integrals U_{ss} are taken to be equal and determined from experiment. The β_{st} are taken to be zero except for nearest neighbours and then their values are given as a function of internuclear distance. An exponential form is often used. The Coulomb integrals are also fixed from experiment. In particular the value of the self-energy integral γ_{ss} is very significant. Its value when estimated from the atomic orbitals is far too large and its reduced value, found by equating it to the difference between the experimental atomic ionization potential and the electron affinity, is essential to achieve reasonable results. This can be interpreted physically as an intra-orbital correlation effect.

A5-3 The Anderson Hamiltonian

Anderson [7] was concerned with the problem of iron group ions alloyed with non-magnetic metals which can produce ferromagnetic behaviour under certain conditions. He used physical arguments to set up his Hamiltonian. There is a term to represent the valence band of the metal, another to add the contribution from the impurity ion, which will depend on its charge, and an interaction term to allow for the transfer of charge from the band to or from the ion. The result takes the form:

$$H = \sum_{k\sigma} \varepsilon_k n_{k\sigma} + E(n_{d+} + n_{d-}) + U n_{d+} n_{d-} + \sum_{k\sigma} V_{dk}(c_{k\sigma}{}^\dagger c_{d\sigma} + c_{d\sigma}{}^\dagger c_{k\sigma})$$

where ε_k is the energy of an electron in the band which has momentum k and $n_{k\sigma}$ is the population of that state with spin σ. The population of electrons of spin $+$ on the impurity is n_{d+} and n_{d-} is the population with $-$ spin. The atomic energy, which is proportional to its total charge, is the second term and U allows for the repulsion between electrons on the same atom. The final term involves the creation operators $c_{k\sigma}{}^\dagger$ and $c_{d\sigma}{}^\dagger$ for electrons in the k band state and the atomic d state respectively along with the corresponding annihilation operators. (For a discussion of these operators see Appendix 4.) The term in U is the major novelty. It recognises the importance of including this strong Coulomb repulsion. The interaction potential V is also developed further into local Coulomb contributions, that from nearest neighbours of the impurity being the largest.

With this model Hamiltonian, Anderson was able to show that these materials would undergo a transition from doubly occupied orbitals, and diamagnetism, to singly occupied orbitals, and ferromagnetism, at a certain critical value of U.

The Anderson model has been reinterpreted as a one-dimensional model of chemisorption by Grimley and Walker [8]. They take the impurity as an adatom at one end of a metal chain. Their analysis shows how an atom can be bound to the surface of the metal through a surface state of the metal.

Another example of the use of a model of chemisorption in which the effect of an adsorbed atom is considered along with a transfer of a fractional charge to or from the body of the material is given by Nakatsuji [9]

A5-4 The Hubbard model

The Hubbard model [10] can be considered as a simplification of the Anderson model. It was introduced to deal with pure ferromagnetic materials. It considers only the atoms with their localized d shell electrons. The model allows for one such orbital per atom and one electron occupying it. Thus the interaction with the valence band has disappeared. The Hamiltonian is written in operator form as:

$$H = \sum_{ij\sigma} T_{ij} c_{i\sigma}{}' c_{j\sigma} + \frac{1}{2} U \sum_{i\sigma} n_{i\sigma} n_{i-\sigma}$$

where i and j label the various atoms. The matrix T_{ij} has diagonal terms for the energy of each atomic orbital and hopping terms for all nearest neighbours. It is easily seen that, for alternant molecules which have unit charges and zero next-neighbour bond orders, this is a good approximation to the PPP equations since the resonance integral can be redefined to include an average bond order term:

$$T_{12} = \beta - \frac{1}{2}P_{12}\,\gamma_{12}.$$

Hubbard shows that correlation effects will introduce a screening of the two-electron integrals and uses this argument to justify his use of a reduced value of U.

The significance of these models is that the stability of the ferromagnetic state is determined by the value of the integral U. By the repulsion between the electrons when an atomic orbital is occupied by electrons of opposite spin the usual ground state which involves doubly occupied orbitals is made less stable. Clearly the value of U is much larger in the inner d shell of the atom than it is in the π orbitals. Thus, this is the origin of the Heisenberg "exchange" term which is introduced into the theory of the cooperative effect.

References

[1] Pariser R 1990 Intern J Quant Chem 37: 319; Parr R G 1990 Intern J Quant Chem 37: 327; Pople J A 1990 Intern J Quant Chem 37: 349 and later papers in the same issue
[2] Lennard-Jones Sir J E 1949 Proc Roy Soc A198: 1, 14
[3] Hall G G 1951 Proc Roy Soc A205: 541
[4] Roothaan C C J 1951 Rev Mod Phys 23: 61
[5] Löwdin P O 1950 J Chem Phys 18: 365
[6] Hall G G 1952 Proc Roy Soc A213: 102; 1954 Trans Faraday Soc 50: 773
[7] Anderson P W 1961 Phys Rev 124: 41
[8] Grimley T B and Walker S M 1969 Surf Sci 14: 395
[9] Nakatsuji H 1987 J Chem Phys 87: 4995
[10] Hubbard J 1963 Proc Roy Soc A276: 238

Appendix 6 The LEPS interaction

A6-1 Valence bond theory

The first attempt to calculate the electronic energy surface for the interaction between the atoms involved in a substitution reaction was made by London [1] using the valence bond theory. He showed that the lowest states of H_3 were doublets and, if the three atomic orbitals for the three atoms were a, b, c, then their energies were given as

$$E_\pm = Q_{ab} + Q_{bc} + Q_{ca}$$
$$\pm \frac{1}{\sqrt{2}} \left((A_{ab} - A_{bc})^2 + (A_{bc} - A_{ca})^2 + (A_{ca} - A_{ab})^2 \right)^{1/2}$$

where the Q_{ab} etc. are the diatomic Coulomb integrals and the A_{ab} etc. are exchange integrals. This form is symmetrical in the three atoms and merges into the correct expressions, in valence bond theory, for the two energy states of the remaining diatomic as any one atom is removed.

A6-2 Use of Morse potentials

The advantage of this theoretical expression in describing correctly the limiting cases was appreciated by Eyring and Polanyi [2] but they preferred to use it semi-empirically. The Coulomb integrals became Morse potentials

$$Q(R) = D\{e^{-2\alpha[R-R_e]} - 2e^{-\alpha[R-R_e]}\}$$

since these are easy to use and give a good approximate form to the diatomic potential energies with the correct dissociation energy D and internuclear distance R_e. The exchange integrals were represented as the corresponding anti-Morse functions:

$$A(R) = D\{e^{-2\alpha[R-R_e]} + 2e^{-\alpha[R-R_e]}\}$$

With these interpretations the reaction surface could be predicted from diatomic experimental data on D, R_e and α. These results were later improved by Sato [3] who inserted a factor to simulate the effect of atomic orbital overlap. The energies are now

$E'_\pm = E_\pm /(1 \pm k)$

where k is a constant independent of R. This is the LEPS function.

References

[1] London F 1929 Z Electrochem 35:552
[2] Eyring H and Polanyi M 1931 Z Phys Chem B12:279
[3] Sato S 1955 J Chem Phys 23:592, 2465

Appendix 7 Green's functions

A7-1 Introduction

There are many situations in solid state theory where the use of the Green's function formalism provides a compact and practical solution to a problem. To illustrate the idea of such a function and give one example which is relevant to our problems the derivation of one Green's function, in both the time-dependent and time-independent senses, will be given here. The general principle behind all such functions is discussed briefly. The particular example of a band system is considered since this is relevant to many of the problems of the solid state.

A7-2 Initial conditions on a long chain

We have already, in Appendix 2, considered the solution of the eigenvalue problem for a long chain system. We now consider the consequent problem of the time dependent problem of how the system evolves when started by exciting one particular site.

We take the linear system which has nearest neighbour interactions β alone and, by adjusting the origin of energy, $\alpha = 0$. It will have eigenvector equations:

$$\beta(a_{r-1} + a_{r+1}) = \varepsilon\, a_r$$

and this has the constant phase solution:

$$a_r = A\, e^{ir\theta} \; ; \; \varepsilon = 2\beta \cos\theta.$$

The time dependent equations are:

$$\beta(u_{r-1} + u_{r+1}) = i\frac{\partial u_r}{\partial t}$$

with the initial condition that only one of these u_r is non-zero at t=0. If this special one is selected as the origin for the purpose of counting and its initial value is 1 we get:

$$u_r(0) = \delta_{r0}.$$

It is easily seen that the time dependent eigensolutions are:

$$u_r = A\, e^{ir\theta}\, e^{-i\varepsilon t}$$

for each value of θ. The general solution is found by superimposing these eigensolutions with arbitrary coefficients. This gives:

$$u_r(t) = \int_0^{2\pi} A(\theta)\, e^{ir\theta - 2i\beta t \cos\theta}\, d\theta$$

where $A(\theta)$ is determined by the initial conditions and the choice $A = 1/(2\pi)$ is readily shown to be sufficient. The Bessel function J_n is defined by a similar integral i.e.

$$J_n(z) = \frac{i^{-n}}{2\pi} \int_0^{2\pi} e^{iz\cos\theta + in\theta}\, d\theta$$

Thus we conclude that the solution is:

$$u_r(t) = i^r\, J_r(2\beta t).$$

This is called the time-dependent Green's function for the chain. It is easily seen that all the initial conditions are now satisfied since all the Bessel functions vanish at the origin except $J_0(0) = 1$. Since $J_n(t) = J_{-n}(t)$ the solution is symmetrical around the starting point of the excitation.

It is pleasing that this problem can be solved using a familiar function. From graphs of the Bessel functions in terms of their order and variable, e.g. Jahnke and Emde [1] p.126, a clear idea can be gained of the gradual dispersion of the excitation over the whole chain. The problem of the dispersion of vibrational energy from a point of initial displacement in a chain is mathematically very similar to this problem. It has been treated by Lennard-Jones [2].

This may seem a rather special problem but it has great significance. The more general problem of arbitrary initial conditions over the whole chain can now be solved. The solution is found by superimposing solutions like the one above for each initial displacement. Thus if, at t=0, we have $u_r = f(r)$ then the general solution is:

$$u_r(t) = \sum_s f(s) \, i^{r-s} \, J_{r-s}(2\beta t).$$

This ability to produce the general solution simply is an important property of the Green's function.

A7-3 Time-independent Green's operators

The basic idea behind the time-independent Green's function can be explained most easily in terms of finite matrices. We start from the eigenvalue equation:

$Mv = \mu v,$

where M is a matrix, μ one of its eigenvalues and v the corresponding eigenvector. The Green's matrix is then defined as the inverse:

$G(\lambda) = (\lambda I - M)^{-1}$

where λ is some constant and I is the unit matrix. If λ is not equal to any eigenvalue this inversion of the matrix is possible.

The advantage of this new matrix is seen when some problems are considered. We note that the eigenvalues of M can be defined as the singularities of the matrix $G(\lambda)$ as λ is varied. If the original matrix is modified by adding a perturbation V:

$N = M + V$

then this new matrix will also have a Green's matrix $H(\lambda)$ with:

$H = (\lambda I - M - V)^{-1}$

so that

$(\lambda I - M)H - VH = I$

When this is multiplied on the left by G the result is the equation:

$H - GVH = G$, or $H = G + GVH$.

From this equation (known as the Dyson equation), which is in closed form, various perturbation expansions can be generated. Thus, by simple recursion, H can be expanded in terms of G as:

$H = G + GVG + GVGVG + GVGVGVG \ldots$

This takes the more familiar form of a perturbation expansion when the matrix is written explicitly as

$$G = \sum_i \frac{v_i v_i^*}{\lambda - m_i}$$

where the sum is over all the eigenvectors v_i of M with their eigenvalues m_i and their Hermitean conjugates v_i^*. In this expression the dyadic product $v_i v_i^*$ means that each component of the first vector is multiplied in turn by each component of the second to form a matrix. Dyadic products and matrix functions are described in [3].

When the matrix becomes singular a special step is needed to avoid the definition becoming impossible. This takes advantage of the fact that the matrices are Hermitean and so have real eigenvalues. By taking the value of λ complex in the neighbourhood of the singularity the divergence can be avoided. An example of this is detailed below.

A7-4 Green's matrix for a band

The example of a linear periodic system has been discussed in Appendix 2. It can be used to illustrate the problems and opportunities of the Green's function method when the eigenvalue spectrum is continuous. From that discussion we know that the eigenvectors are defined by:

$$v_n(\theta) = ce^{in\theta}, c = 1/\sqrt{(2\pi N)},$$

with the eigenvalues:

$$m(\theta) = a + 2b\cos\theta, -\pi < \theta < \pi.$$

The Green's matrix is then given by:

$$G_{mn} = \frac{1}{2\pi} \int_{-\pi}^{\pi} \frac{e^{i(m-n)\theta}}{\lambda - a - 2b\cos\theta} \, d\theta$$

We note that G_{mn} is a symmetrical circulant since its elements depend only on the difference of the subscripts. It can be written as G_r where $r = |m-n|$ and this justifies the use of the term function for an entity which is properly defined as an operator. It is usual to omit the factor of N since this is proportional to the volume and is allowed for elsewhere.

The evaluation of this integral can be performed most easily using a contour integral. We consider the following integral over the complex variable z:

$$I_r = \frac{1}{2\pi} \int \frac{z^r}{\lambda - a - bz - b/z} \frac{dz}{z}$$

where the integral is taken on the unit circle around the origin. This gives $z = e^{i\theta}$ and the integral becomes

$$I_r = \frac{i}{2\pi} \int_{-\pi}^{\pi} \frac{e^{ir\theta}}{\lambda - a - 2bc\cos\theta} d\theta$$

$$= i\, G_r.$$

The value of this integral is given by $2\pi i$ times the residues at the poles of the integrand which are inside this circle. Care is needed to consider all the special cases. It is convenient to define, as a new variable, the scaled energy:

$$t = \frac{\lambda - a}{2b}$$

so that the denominator in the contour integral can be factorized as

$$b(z - z_+)(z - z_-)$$

where

$$z_\pm = t \pm \sqrt{t^2 - 1}, \quad |t| > 1$$

and the condition implies that λ is outside the band. Now the product of these two roots is 1 so that one is inside and one outside the unit circle. The one inside, if $t > 0$, is z_- while, if $t < 0$, it is z_+. It is useful to call the inside one z_i.

$$z_i = z_-, \quad t > 0$$

$$= z_+, \quad t < 0.$$

The residue at z_i is then

$$R = \frac{z_i{}^r}{b\sqrt{t^2 - 1}}$$

and the Green's function is now

$$G_r = \frac{z_i{}^r}{2b\sqrt{t^2 - 1}}.$$

When λ is inside the band then $|t| < 1$ and the integral becomes more difficult. The two roots now both lie on the unit circle itself

$$z_\pm = t \pm i\sqrt{1 - t^2}.$$

We avoid this by changing the value of λ, which has been assumed above to be real, so that it has a small imaginary part:

$\lambda \rightarrow \lambda+i\varepsilon$, $\varepsilon>0$.

This has the effect of making t complex, of moving the upper root outside the circle and the lower one inside so that

$$z_i = t - i\sqrt{1-t^2}$$

and the residue is taken at this point. This gives:

$$R = \frac{\left(t - i\sqrt{1-t^2}\right)^r}{2b\sqrt{1-t^2}}$$

and, if we define γ by

$$e^{i\gamma} = t + i\sqrt{1-t^2},$$

then the Green's function is:

$$G_r = -i\frac{e^{-ir\gamma}}{2b\sqrt{1-t^2}} \quad \text{(inward solution)}$$

Alternatively, if we use t-iε, we obtain the solution:

$$G_r = i\frac{e^{ir\gamma}}{2b\sqrt{1-t^2}} \quad \text{(outward solution)}.$$

where the two solutions correspond to inward and outward waves respectively.

One advantage of this formalism is that the density of states can be deduced immediately from the imaginary part of the Green's matrix:

$$\rho(\mu) = -\pi^{-1} \text{ Im Tr } G_{mn}(\mu)$$

where μ is inside the band and Tr denotes the operation of taking the trace of the following matrix. But all the N diagonal elements of this matrix are equal so, omitting the N factor, this reduces to:

$$\rho(\mu) = -\pi^{-1} \text{ Im } G_0(\mu)$$

$$= \frac{1}{2\pi b\sqrt{1-t^2}} = \frac{1}{\pi\sqrt{4b^2-(\mu-a)^2}}$$

This agrees with the density derived from first principles in Appendix 2. It is easy to check that it integrates to unity over the band. The angle γ also is equal to the angle θ defined in Appendix 3 as the phase change at an impurity with λ here replacing A there.

This may seem to be a different problem from the time-dependent problem in the first section above. Nevertheless there is a connection.

When the Bessel functions are subjected to a Fourier transform they turn into the expressions here. In general the relation between the time-dependent and time-independent Green's operators is always equally close since:

$$G_r(\varepsilon) = -i \int_{-\infty}^{0} u_r(t)\, e^{i\varepsilon t}\, dt \quad \text{Im } \varepsilon < 0$$

where $u_r(t)$ is the time-dependent Green's operator.

Further extensions of this Green's function technique using two components in each cell have been given by Ueba [4].

References

[1] Jahnke E and Emde F 1945 Tables of Functions, Dover New York.
[2] Lennard-Jones J E 1937 Proc Roy Soc A163: 127
[3] Hall G G 1963 Matrices and Tensors, Pergamon London
[4] Ueba H 1980 Phys Stat Sol B99: 763

Appendix 8 Atomic units

A8-1 Theoretical units

Experimental measurements are based on some units which can be defined with as great precision as is required but may be selected arbitrarily. Since the discovery of the electron, theoretical calculations on electronic properties of atoms and molecules can be related to the properties of the electron itself. This provides a natural system of units, atomic units. It has the major advantage that the calculations can be performed without regard to any experimental measurement. When the mass of the electron, for example, is revised through more exact experiment the final conversion factor which relates the calculation to experiment will need revision but the calculation itself will be unaffected.

The system of atomic units that is used most frequently (there are two other systems!) was devised by Hartree [1]. He took the electron mass, m, as the mass unit; the unit of charge, e, is the negative of the electron charge and the quantum constant, $\hbar$, is the unit of action. These three units determine all the others required for quantum calculations. In particular the unit of length becomes the first Bohr radius:
$$a_0 = \hbar^2/me^2$$
which is now known as the **Bohr** (Shull and Hall [2]). The unit of energy is:
$$H = e^2/a_0$$
which is called the **Hartree** in honour of the originator of these units.

The unit of time is defined in terms of these units as
$$\tau = \hbar^3/me^4$$
and has the approximate value of 2.4 10^{-17} sec. From these the unit of velocity becomes:
$$a_0/\tau$$

and is approximately 2.2 10^8 cm sec^{-1}. In these units the velocity of light is $1/\alpha$, where α is known as the fine structure constant:

$\alpha = e^2/c\hbar$

with the value 1/137.036. Another important quantity for some calculations is the proton mass, M. The ratio M/m is involved in the coupling of electronic and nuclear motions.

A8-2 Experimental values

The accepted values [3] of these units relative to conventional experimental units at the date, 1988, are given below together with some conversion factors.

mass   m   9.10939 10^{-31} kg
charge   e   1.602177 10^{-19} C
action $\hbar$ 1.0545731$0^{-34}$ J sec
length a_0 0.529177 10^{-10} m

fine structure constant α =1/137.036

energy H 4.359748 10^{-18} J

proton M/m = 1836.1527

It is convenient to have the equivalence that H corresponds to 27.211396 eV and to 627.51 kcal/mol.

For convenience we also include the units required in the thermodynamic calculations. These start from the number of molecules in a gram molecule N_0 = 6.0221367 10^{23}. The Boltzmann constant is then

k   1.380658 10^{-16} erg/deg

and the gas constant

R   8.31451 J/deg mole = 1.98 kcal/deg mole.

References

[1] Hartree D R 1926 Proc Camb Phil Soc 24: 89
[2] Shull H and Hall G G 1959 Nature 184: 1559
[3] Cohen E R and Taylor B N 1988 J Phys Chem Ref Data 17: 1795

Author Index

Subject Index

P. R. Surján, Budapest

Second Quantized Approach to Quantum Chemistry
An Elementary Introduction

1989. XIII, 184 pp. 11 figs. 1 Tab. Hardcover DM 128,- ISBN 3-540-51137-7

Contents: Introduction. - Concept of Creation and Annihilation Operators. - Particle Number Operators. - Second Quantized Representation of Quantum Mechanical Operators. - Evaluation of Matrix Elements. - Advantages of Second Quantization. - Illustrative Examples. - Density Matrices. - Connection to "Bra and Ket" Formalism. - Using Spatial Orbitals. - Some Model Hamiltonians in Second Quantized Form. - The Brillouin Theorem. - Many-Body Perturbation Theory. - Second Quantization for Nonorthogonal Orbitals. - Second Quantization and Hellmann-Feynman Theorem. - Intermolecular Interactions. - Quasiparticle Transformations. Miscellaneous Topics Related to Second Quantization. - Problem Solutions. - References. - Index.

The subject of this book is the application of the second quantized approach to quantum chemistry. Second quantization is an alternative tool for dealing with many-electron theory. The vast majority of quantum chemical problems are more easily treated using second quantization as a language. This book offers a simple and pedagogical presentation of the theory and some applications. The reader is not supposed to be trained in higher mathematics, though familiarity with elementary quantum mechanics and quantum chemistry is assumed. Besides the basic formalism and standard illustrative applications, some recent topics of quantum chemistry are reviewed in some detail. This book bridges the gap between sophisticated quantum theory and practical quantum chemistry.

Springer-Verlag
Berlin
Heidelberg
New York
London
Paris
Tokyo
Hong Kong
Barcelona

R. L. Carlin, University of Illinois, IL

Magnetochemistry

1986. XI, 328 pp. 244 figs. 21 tabs. Hardcover DM 118,–
ISBN 3-540-15816-2

Contents: Diamagnetism and Paramagnetism. – Paramagnetism:
Zero-Field Splittings. – Thermodynamics. – Paramagnetism and
Crystalline Fields: The Iron Series Ions. – Introduction to Magnetic
Exchange: Dimers and Clusters. – Long-Range Order. Ferro-
magnetism and Antiferromagnetism. – Lower Dimensional Mag-
netism. – The Heavy Transition Metals. – The Rare Earths or
Lanthanides Ions. – Selected Examples. – Some Experimental
Techniques. – Formula Index. – Subject Index.

Professor Carlin's book integrates crystal field theory and the theory
of the electronic structure of transition metal complexes with their
paramagnetic properties. It goes on to consider the properties of inter-
acting magnetic ions, culminating in a discussion of magnetic ordering phenomena. Illustrations of low dimensional ordering are
included, as well as a systematic survey of the literature. The book
brings up to date Professor Carlin's earlier book on "Magnetic Properties of Transition Metal Compounds" (Inorganic Chemistry
Concepts, Vol. 2, Springer-Verlag 1977), and
also provides more introductory material.
A chapter on experimental procedures has
been added, as well as one on the heavier
transition metals and another on the
lanthanides. The book is unique in its expo-
sition of magnetic ordering phenomena,
field dependent properties and its synthesis
of magneto-structural correlations. It is
written as a text, but researchers will find
the literature survey also of great value.

Springer-Verlag
Berlin
Heidelberg
New York
London
Paris
Tokyo
Hong Kong
Barcelona